An Introduction to Health and Safety Law

A student reference

David Branson

Routledge
Taylor & Francis Group

LONDON AND NEW YORK

First published 2015
by Routledge
2 Park Square, Milton Park, Abingdon, Oxon OX14 4RN

and by Routledge
711 Third Avenue, New York, NY 10017

Routledge is an imprint of the Taylor & Francis Group, an informa business

© 2015 David Branson

The right of David Branson to be identified as the author of this
work has been asserted by him in accordance with sections 77 and
78 of the Copyright, Designs and Patents Act 1988.

All rights reserved. No part of this book may be reprinted or
reproduced or utilised in any form or by any electronic, mechanical,
or other means, now known or hereafter invented, including
photocopying and recording, or in any information storage or
retrieval system, without permission in writing from the publishers.

Trademark notice: Product or corporate names may be trademarks
or registered trademarks, and are used only for identification and
explanation without intent to infringe.

British Library Cataloguing-in-Publication Data
A catalogue record for this book is available from the British Library

Library of Congress Cataloging-in-Publication Data
Branson, David (Lawyer)
 An introduction to health and safety law: a student reference/
David Branson.
 pages cm
 Includes bibliographical references and index.
 1. Industrial safety – Law and legislation – Great Britain.
 2. Industrial hygiene – Law and legislation – Great Britain.
 3. Great Britain. Health and Safety at Work etc Act 1974.
 4. Great Britain. Corporate Manslaughter and Corporate
Homicide Act 2007. I. Title.
 KD3168.B73 2015
 344.4104'65 – dc23
 2014023142

ISBN: 978-1-138-01843-3 (pbk)
ISBN: 978-1-315-77117-5 (ebk)

Typeset in Sabon
by Florence Production Ltd, Stoodleigh, Devon, UK

Printed and bound by CPI Group (UK) Ltd, Croydon, CR0 4YY

Contents

Acknowledgements

I would like to thank various people for their help to me in writing this book. First my wife, who has been very supportive during the whole enterprise and has always been there to provide encouragement. Second I would like to thank my sister-in-law Margaret Joy, for all her work in proof reading, a task which is so important and yet often not properly appreciated.

In addition, I would like to thank all the staff at Middlesbrough College who have given me support, including Carol Stephenson, my fellow law lecturer, who advised me on the area of legal practice; and also the library staff who were so patient in waiting for me to return the books I had borrowed. I am also grateful for the support of the staff at Evershed's, whose law update sessions I found so enjoyable and informative. Finally, I would like to thank all the students I have taught over the past 18 years on various NEBOSH courses, whose enthusiasm and interest in the subject have been a great encouragement to me. I hope they find this book both informative and enlightening.

David Branson
May 2014

Table of cases

Cases that appear in the appendix are in *italics*.

EU Cases

Table of statutes

Table of regulations and orders

Introduction

The problem with health and safety law

In recent years, the area of health and safety law has become a political football. The law is seen as too restrictive, preventing the running of donkey races on the beach, creating excessive beaurocracy and stifling the ability of managers to manage their organisations effectively. As a result, the report of Lord Young, entitled 'Common Sense; Common Safety', was commissioned by the incoming Coalition Government with the task of trying to simplify the system and remove excessive health and safety regulations.[1] In all, health and safety has come to be seen as a stumbling block to management efficiency and even to our enjoyment of life.

In looking at this problem, we also need to be aware of the changing nature of society and the world of work in the United Kingdom. We live in a society that has conquered many of the illnesses that used to kill off people at an early age. This is reflected in the rapid rise of life expectancy from 42 at the beginning of the twentieth century, to a figure now approaching the low 80s. In addition, the number of people killed at work has dramatically fallen from over 600 a year in 1974 to around 200 in recent years. This however does not compare to the fall in workplace fatalities since the beginning of the twentieth century, for which figures are less easy to obtain. It is clear that the chance of being killed at work is now seen as very remote today, while in our great grandfathers' day, death at work was a very real possibility. To a large extent, this fall is due to the gradual change from a manufacturing to a service economy, but it is also the result of changing management practices, which are a result of a much greater intolerance of the public towards the risk of death or injury at work.

To some extent, this is reflected in our attitude to safety in the wider world. We do not expect to be killed or injured while going about our everyday activities and we are much less willing to put up with injuries without seeking someone to blame. The increasing number of civil legal actions against private employers and public authorities reflects this change in attitude. The availability of no win no fee legal services has helped to promote this trend, with people now unwilling to put accidents down to bad luck, but instead seeking to pin liability on another person or organisation and to obtain compensation.

It is no surprise that the above trends have been reflected in the development of criminal and civil legal liability in the area of health and safety. The greater reluctance to accept workplace accidents as a fact of life has meant that employers are more likely to be sued by injured employees and the courts are more likely to award compensation. Indeed there is a developing view that a person who is injured at work should be able to claim, whether or not the employer is actually at fault. This can be seen in the seminal case of *Stark* v *Post Office (2000)*, where **damages** were awarded in the case of a failure to maintain work equipment, when in fact no reasonable employer could have been expected to discover the fault. This case was decided by reference to the *Provision and Use of Work Equipment Regulations 1998*, which imposed **strict liability** on employers. The decision reflects the views of many people: that the employer should be liable in such situations, rather than the loss falling on the employee. Nevertheless, the right to take action for breach of statutory duty has recently been abolished by the *Enterprise and Regulatory Reform Act 2013* following pressure from employers to limit the liability of employers for no-fault civil liability.

The development of criminal liability has often been in reaction to changing public opinion. The severe loss of life on Piper Alpha in 1988 was a factor that led to the introduction of more stringent controls over the operation of oil and gas platforms. Similarly, the failure to mount a successful **prosecution** of the management of P&O Ferries in 1991, after the similarly large loss of life with the sinking of the Herald of Free Enterprise, ultimately led to the introduction of Corporate Manslaughter legislation by way of the *Corporate Manslaughter and Corporate Homicide Act 2007*. These are just two examples of legislation following on from a disaster that had impact on public opinion, but there are many more.

However, we also need to be aware of an undercurrent in economic thinking, which is arguably inimical to health and safety concerns. The growth of the globalised economy has meant that the UK now finds itself in competition with developing economies, such as China or other countries in the Far East, where there is no tradition of health and safety legislation or the political will to introduce such protection. There is a growing feeling among some in business that we have to relax our health and safety controls in order to compete effectively with these emergent economies. In addition, the employment relationship has changed, with some employers seeking to engage staff on new forms of short-term and agency contracts, in order to minimise their legal liabilities. There is a concern that this may become a means of undermining health and safety protection.

In this respect, it is important to note the recent concerns over the impact of European Law on health and safety legislation. Under Article 153 of the Treaty of Rome (as amended), the European Union has an obligation to ensure the health and safety of workers by means of setting minimum health and safety standards. This has been translated into domestic legislation such as the *Working Time Regulations 1998*, which limit working hours for employees to an average of 48 hours a week. In the UK, there has been much opposition to this from some employer groups, who see it as unacceptably restricting their ability to compete in the international market place. In this way, health and safety has been drawn into a much wider political arena dealing with the issue of international competitiveness.

The structure of the book

The aim of this book is to provide a brief overview of health and safety law in the UK. As such, it is designed to provide students on the NEBOSH course with a better understanding of health and safety law, as well as addressing some of the key legal issues that have arisen in recent years. The area of health and safety law is often misunderstood by health and safety prac-titioners, who tend to be practical people, not always used to legal concepts or arguments. I hope that this book will clarify the main points at issue, instead of complicating matters. For this reason, I have opted to simplify matters as far as possible by avoiding too much detail. Students who wish to obtain a much deeper understanding of the law, should consult the more specialist books available such as *The Law of Health and Safety at Work* by Norman Selwyn.

In this book, I would like to start in Chapter 1 by looking at the legal structure in the UK, including the sources of law and the nature of the judicial system. In Chapter 2, I will move on to look at the basis of criminal liability for health and safety law, focussing on the *Health and Safety at Work Act 1974* and relating it to recent legal developments. I will follow this, in Chapter 3, by looking at civil liability, starting with the **Tort** of Negligence, before discussing the possible actions for breach of statutory duty. In Chapter 4, I will discuss contractual liability and the impact of Employment Law on health and safety, as well as looking at remedies in the Tort of Nuisance. In Chapter 5, the key subordinate legislation will be covered, looking at the recent case law developments in key regulations, such as the *Provision and Use of Work Equipment Regulations 1998* and the *Workplace (Health, Safety and Welfare) Regulations 1992*. Chapter 6 will look at enforcement arrangements for Criminal Law and Chapter 7 will look at the remedies available in Civil Law. Finally, Chapter 8 will provide a brief conclusion, focussing on the impact of the recent developments in health and safety law. The appendix is a summary of the key legal cases in this area of law, with a commentary on their importance.

I trust that this approach will serve the purpose of providing easy-to-access source material. Yet, at the same time, I will discuss some of the key legal issues that have become prominent in recent times. It is important to look at these areas, because only then are we able to get some idea of how this area of law is likely to develop in the future. The law of health and safety is in flux at present, and we are likely to see some changes due to rising concerns over the impact of litigation culture if nothing else. Yet we must be wary of undermining the positive achievements of the system as a whole. The UK is a world leader in health and safety and this is one of the reasons why UK Health and Safety qualifications are now being adopted worldwide. We have an enviable record in reducing death and injury at work, and this is one that we must endeavour to maintain.

Note

1   'Common sense and common safety', report by Lord Young on health and safety law

1 The legal framework

Introduction

In this chapter, I would like to look at the basic legal structure in the United Kingdom, as it affects the law of health and safety. I will start by looking at the difference between Criminal and Civil Law, before discussing the sources of law and the adjudication system. In Chapter 6 I will discuss the enforcement system for criminal liability as well as the role of the regulatory authorities, such as the Health and Safety Executive (HSE) and the local authorities, while in Chapter 7 I will look at the nature of the civil remedies available.

The types of law

Most legal systems, including that of the UK, are divided into Criminal Law and Civil Law systems. These two areas of law are very different in their nature, having a different purpose, different methods of adjudication and a different outcome. The problem is that these areas of law are inherently linked in health and safety, thereby leading to considerable confusion. The *Lofstedt Report*[1] into health and safety law proposed a number of changes, including, most controversially, the ending of the right to sue for breach of the strict liability statutory duties as implemented by the *Enterprise and Regulatory Reform Act 2013*. This will mean that some of the overlap between Civil and Criminal Law will now be of less importance, although it remains to be seen how the courts interpret the Common Law liability in the light of this change.

Criminal Law

The purpose of Criminal Law is to protect the public from harm, and that is why the state, through its various agencies, takes such a prominent role in the criminal process, not only providing the court structure, but also initiating the legal process by undertaking a prosecution. The underpinning aim here is deterrence, imposing punishments on those parties who breach the law in order to dissuade others from following suit. The key role of publicity is important, and the HSE will put convictions on their website to help bring this information to the notice of other parties. It is also the case that a criminal conviction may make it difficult for a party to obtain future contracts, as there is the possibility that they may re-offend and, in that case, implicate the party they work for. In this way, the effect of a criminal prosecution is quite significant and businesses will strive to avoid a prosecution where possible.

In health and safety cases, the prosecution is usually initiated by the HSE or the local authority, or in the case of prosecutions for **manslaughter**, by the Department of Public Prosecutions (DPP). The case is brought to ensure that the standards of health and safety are maintained, so there is no requirement for a party to have suffered injury for a case to be brought.

Criminal Law in health and safety

GENERAL LIABILITY

In health and safety law, criminal liability itself falls into two main categories. First, there is a general liability under the *Health and Safety at Work etc. Act 1974 (HSWA)*, and in respect of death at work under the *Corporate Manslaughter and Corporate Homicide Act 2007 (CMCHA)*. Both of these pieces of legislation impose liability as regards the failure of the organisation to manage health and safety effectively.

Under Section 2 of *HSWA* the employer has a duty to ensure 'the health, safety and welfare at work of his employees'; while under Section 3 the employer has to 'conduct his undertaking in such as way as to ensure . . . that persons not in his employment . . . are not thereby exposed to risks to their health and safety'. The focus here is on how the enterprise is managed to ensure the health and safety of employees and this general duty is to some extent further explained in Section 2(2) of the act. Here the duties are: to provide and maintain safe plant and systems of work; to provide adequate information, training and supervision; to maintain a safe place of work; and to provide and maintain a safe working environment. These duties are based on the Common Law duties laid down in *Wilsons and Clyde Coal Co Ltd v English (1938)* and are very general in scope.

The other generalised area of criminal liability is the law on manslaughter, which may impose both an individual or corporate liability. Previously this involved liability at common law, but this has been supplemented by statutory liability under *CMCHA*. This legislation imposes a liability on any organisation that fails to effectively manage health and safety, resulting in the death of any person to whom they owe a duty of care. Like *HSWA*, the legislation focuses on the management of health and safety and is very general in nature. However, unlike *HSWA*, it is result-driven, in so far as liability is determined by the consequences of the breach. Therefore there is no liability under the act if there is no fatality, even if the end result is serious injury or no injury at all. This is quite different from *HSWA* or the subordinate legislation, where liability is based on the failure of the organisation to manage health and safety itself, independent of the consequences.

SPECIFIC LIABILITY

This generalised liability is supplemented by a wide range of subordinate legislation, usually passed under Section 15(1) of *HSWA*. These cover a wide variety of situations broadly based on specific hazards. So, for example, machinery safety is covered by the *Provision and Use of Work Equipment Regulations 1998 (PUWER)*; safety of the workplace is covered by the *Workplace (Health, Safety and Welfare) Regulations 1992 (WHSWR)* and the problem of noise by the *Control of Noise at Work Regulations 2005 (CNWR)*. These regulations impose liabilities that may be strict, as under Regulation 5 of *PUWER*, or may be qualified by the term 'reasonably practicable', as in Regulation 12(3) of *WHSWR*.

Civil Law in health and safety

The purpose of Civil Law is quite different from Criminal Law. Here, the main aim is to provide a remedy for persons who have suffered a loss, such as being injured at work. The legal action is designed to enable the **claimant** or his dependants to obtain financial compensation from the **defendant**. In cases of accidents at work, this is normally paid out of an insurance policy, which employers are required to have by law in respect of their employees, by virtue of the *Employers Liability (Compulsory Insurance) Act 1969*.

In Civil Law, the courts are mainly concerned to ensure that an injured party obtains some redress. This has meant that the courts have been ready to modify the law in order to facilitate this. Over the last few years many of the defences to liability have been restricted, such as **volenti** (or consent) and contributory negligence. In addition, the rules on legal **causation** have been relaxed to allow claimants to pursue actions for compensation in respect of work-related diseases, where it is not really possible to prove a direct causal relationship, due to the limitations of scientific knowledge. This can be seen in the **mesothelioma** cases such as *Fairchild* v *Glenhaven Funeral Services Ltd (2002)*. The position nowadays in law is that it is very difficult for an employer to avoid liability in respect of a claim for an injury caused at work. In effect, we have created a kind of strict liability regime.

GENERAL LIABILITY (NEGLIGENCE)

In Civil Law, there is a general liability under the common law Tort of Negligence (Delict in Scotland), although it is always possible to bring an action for breach of the contract of employment as an employee.

The liability in tort was developed in the nineteenth century in cases such as *Smith* v *Baker & Sons (1891)*, which severely limited the defence of volenti. In addition, the doctrine of common employment, which held that an employer was not liable for the actions of his employees, was effectively abolished by Section 18 of the *Law Reform Act 1936*. Finally, the defence of contributory negligence, which had been a complete defence, was restricted to a partial defence in the *Law Reform (Contributory Negligence) Act 1945*.

The cumulative effect of all these developments was that the Tort of Negligence became the key means for an employee to claim compensation for injury at work. In order to claim in Negligence, it is necessary to prove three key elements of the Tort; the existence of a duty of care, the breach of that duty and an injury arising from the breach. The nature of the duty of care is as set out in *Wilson and Clyde Coal* v *English (1938)*, and this involves the provision of safe plant and equipment, safe systems of work, a safe place of work and competent fellow employees, these being of course the basic elements of the later criminal liability as set out in Section 2(2) of *HSWA*. Later Civil Law cases have widened the duty of care, to include adequate supervision and a duty to provide a safe working environment. The defendant will be in breach if he fails to act as a 'reasonable' employer. This in theory creates a **fault-based** liability, and this is underpinned by the fact that the liability is only applied in respect of any loss that is 'reasonably foreseeable'.

However, as regards personal injury, the above position is modified by the so-called 'egg shell skull' rule. This requires the defendant to take the victim as he finds them, making them liable for the full consequences of any reasonably foreseeable injury, even if the medical consequences are unforeseen. For example, in the case of *Smith* v *Leech Brain & Co. (1962)*, the defendant was liable for a fatal cancer caused to the claimant as a result of a splash of molten metal on his lip. As the injury to the lip was seen as reasonably foreseeable, the defendant was liable for the full consequences of the injury, however unforeseeable it might have been.

GENERAL LIABILITY (CONTRACT)

Although the wide scope of the Tort of Negligence has meant that it has been used preferentially as the means of redress in the case of injury at work, there is always the alternative of an action for breach of contract. Under the Contract of Employment a duty of care equivalent to that in Negligence is implied into the contract. Moreover, the limitation period, or period in which an action can take place under contract is six years, compared to three years in the Tort of Negligence. However, where the injury is latent, such as an industrial disease, an action in Negligence is a better option because the three-year period runs from the time the injury is discovered and the defendant has been identified, a situation that may occur after many years with a disease such as **asbestosis**. Such an extension is not possible in Contract Law, so it cannot be used for this purpose.

The main benefit of an action for breach of contract is that it can take place in the jurisdiction where the contract is deemed to apply, whereas an action in the Tort of Negligence can only be brought in respect of an action in the UK. Therefore it is possible to use an action in Contract Law where the claimant is working outside of the UK, but under a contract made under UK law, as in the case of *Matthews* v *Kuwait Bechtel Corporation (1959)*. However, an action can still only be brought if there is a contract, and this may not always exist.

GENERAL LIABILITY (NUISANCE)

There is also a general liability under the Tort of Nuisance, which mainly deals with damage to property rather than injury to persons. This involves the Torts of Private Nuisance and Public Nuisance, the first of which is essentially a land based Tort, but the latter of which gives rise to actions by a group of persons who suffer special damage, and this can include personal injury. Of particular importance is the specialised action in the tort of *Rylands* v *Fletcher*, which imposes strict liability for the release of dangerous substances onto a person's land, a matter which is of some importance for health and safety.

The recent case of *Cambridge Water Co. Ltd* v *Eastern Counties Leather PLC (1994)* has modified the nature of the liability under the tort of *Rylands* v *Fletcher*, by limiting compensation to damage that is reasonably foresee-able, thereby creating a liability not too different from the Tort of Negligence. Although this area of law is only indirectly related to health and safety, it is important to be aware of the key principles, as there are sometimes benefits in taking an action in Nuisance rather than Negligence. In particular, it is not possible to claim for pure economic loss or loss of profits in Negligence, but this may be possible in Public Nuisance. This may be of importance in the case of those who are self-employed, who may lose the opportunity to obtain future business when they are injured, especially if they cannot sue for breach of contract.

SPECIFIC LIABILITY (BREACH OF STATUTORY DUTY)

In addition to the general liability under the Tort of Negligence, it was possible until recently to take an action for breach of statutory duty, using the subordinate legislation used to impose criminal liability, such as *PUWER* or *WHSWR*. The regulations often allowed for civil as well as criminal liability, and they could be very useful to the claimant because they some-times imposed strict liability, whereas Negligence is based on fault-based liability. In some cases, claimants were able to successfully pursue a claim for breach of a statutory duty when an action in Negligence was not poss-ible. In the case of *Stark* v *Post Office (2000)* the claimant was injured by a defective bicycle where the fault was impossible to detect by normal inspection. As such, the employer could not be liable in Negligence as no 'reasonable' employer could have foreseen the accident. However, he was able to obtain a remedy under Section 6 of the then *PUWER 1992*, which required equipment to be maintained 'in an efficient state, in efficient working order and in good repair'.

The statutory remedy is sometimes strict in nature, whereas the Negligence liability is fault-based. However, strict liability in Civil Law came in for criticism in the *Lofstedt Report*,[1] as imposing an unreasonable burden on employers, which was not intended by government. As a result, civil actions for breach of the health and safety subordinate legislation has now been abolished under the *Enterprise and Regulatory Reform Act 2013* as referred to above. What will be interesting to see, is whether the courts still allow a level of strict liability by means of a wide interpretation of fault-based liability under common law Negligence. In addition, there are some statutes that impose purely civil liability, such as the *Occupiers' Liability Acts 1957* and *1984*, but these only apply to the liability of an employer as an occupier of premises.

We can see that civil liability partly reflects the nature of liability in Criminal Law. In both cases, we have a general liability that is essentially fault-based in nature, based on what is 'reasonably practicable' for Criminal Law or 'reasonably foreseeable' for Civil Law. In addition, it is possible to use the subordinate legislation for criminal actions, although no longer for civil ones, and here the nature of the liability will vary according to the relevant statute.

The sources of law

Just as there are two main types of law for health and safety, there are two main sources of law; namely legislation (or statute) and case law. Although legislation is by far the most important, especially in criminal liability, we should still be aware that both sources of law have a role to play in determining liability. In addition, we should note that both the UK and the EU now have a role in passing legislation that affects health and safety. The sources of law as are as indicated on Figure 1.1.

United Kingdom legislation

Acts of Parliament

Legislation is law passed by Parliament, and as such this includes Acts of Parliament, Regulations and Orders. The key difference between Acts of Parliament and the rest is that Acts of Parliament have to go through a complex procedure in Parliament to become law. This involves three readings in the House of Commons and another three in the House of Lords, before receiving the Royal Assent. This procedure is very time consuming and so

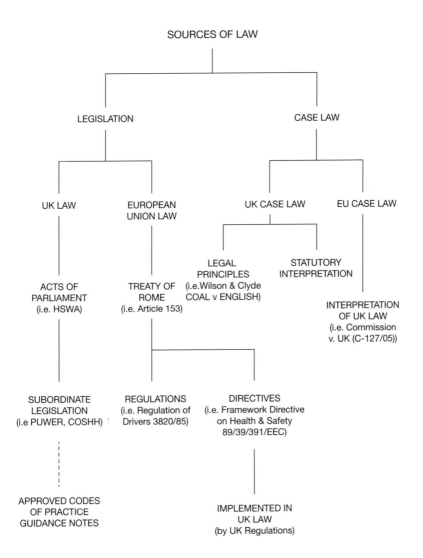

Figure 1.1 Sources of law
Source: Diagram by author.

is reserved for controversial legislation, where the legislators wish to scrutinise the proposed legislation in detail. This was the method used for much earlier legislation such as the Factories Acts. However, in recent times it had been reserved for only a few key acts such as the *Health and Safety at Work Act 1974 (HSWA)* and the *Corporate Manslaughter and Corporate Homicide Act 2007 (CMCHA)*, both of which engendered a great deal of controversy at the time.

The *HSWA* introduced key new proposals such as 'goal setting' instead of prescriptive controls, as well as extending the liability to a much wider range of persons. The Corporate Manslaughter legislation dealt with the problem of affixing corporate liability for manslaughter without having to indentify a **'directing mind'**, an issue that had made it virtually impossible to pin liability on large corporations, as highlighted in the case of *R* v *P & O Ferries (1991)*. It should be noted that Acts of Parliament are the supreme form of legislation and can overrule any decision of the courts in the UK. This can be seen in the way that the *Compensation Act 2006* was swiftly introduced to overrule the decision in the case of *Barker* v *Corus (UK) Ltd (2006)*, which had restricted the ability of claimants to pursue mesothelioma claims against multiple defendants.

Subordinate legislation

In the case of subordinate legislation, such as regulations or orders, these are passed under the provision of an enabling act, such as under Section 15 of the *Health and Safety at Work Act 1974*. This type of legislation is often initiated by the enforcement authorities or a minister, and involves extensive consultation with interested parties before it is introduced to Parliament. Such legislation is passed through Parliament by a much swifter method, which does not give the same opportunities for parliamentary scrutiny. In the case of health and safety legislation, the 'negative resolution' procedure is normally used, by which a piece of subordinate legislation is laid before both Houses of Parliament and will be passed unless there is an objection. However, we should note that the legal position of subordinate legislation is less effective than an Act of Parliament. Although they can overrule judicial decisions, they cannot exceed the legal scope of the enabling act and they can be struck down by the court if they are seen to be unreasonable.

Codes of Practice and guidance notes

Finally, we should note the legal position of the Approved Codes of Practice, often accompanying the legislation. These are passed under Section 16 of the *Health and Safety at work Act 1974* and are designed to provide practical guidance on how to implement the relevant regulations. In this respect, Section 17 of *HSWA* states that while failure to comply with a Code of Practice is not a breach of Civil or Criminal Law in itself, it is admissible in evidence. No reference is made to the position of a Code of Practice in Civil Law, but under Section 11 of the *Criminal Evidence Act 1968* evidence of a criminal conviction is admissible in civil proceedings, so it would have an indirect impact should there be both a criminal and a civil action stemming from the same events. The position of Guidance Notes is less specific, and it would appear that they offer more practical advice, which may or may not be followed. However, it would seem that where a person follows such guidance, they would be unlikely to incur any criminal or civil liability.

European legislation

Institutions of the European Union

INTRODUCTION

The European Union was set up by the *Treaty of Rome 1957* and has expanded from its 6 original members to number 28 members following the recent accessions. The United Kingdom joined in 1973 and is bound by the law of the community under the provisions of the *European Communities Act 1972*. There have been a number of treaties amending the constitution of the EU and the article numbers were amended by the *Treaty of Amsterdam 1997* and later by the *Treaty of Lisbon*. I have used the most recent article numbers but have also referred in brackets to the article numbers under the *Treaty of Amsterdam*, as they are the ones most commonly known.

The objective of the EU is to approximate the economic policies of the member states, in order to ensure a continued and balanced expansion of the community and the raising of living standards. As regards health and safety, the aim is set down in Article 153 (previously Article 137) of the *Treaty of Rome* and involves harmonising the laws relating to working conditions in order to improve the health and safety of workers. The EU is a 'supranational' body and as such constitutes a separate legal order, distinct from the national law of the member states. As such, the EU confers legal rights directly on individual citizens, which can be enforced in the courts of the member states as well as in the European Court of Justice (ECJ).

There are three main institutions of the European Union: the Council of the European Union, the European Commission and the European Court of Justice. The legal relationship between these bodies is regulated by the *Treaty of Rome* and is still in a state of flux, although the general rules have become clearer over time.

THE COUNCIL OF THE EUROPEAN UNION

The Council of the European Union (previously the Council of Ministers) is composed of government ministers from all of the member countries who are delegated to attend for that particular session. The Council is the main policy-making body and is responsible for initiating much of the community legislation. Voting is usually by qualified majority on a weighted voting system, although some issues still require unanimity. Such a system means that the legislation must be supported by the overwhelming majority of the member states, but it is still possible to overrule the objections of a single member state, as was the case with the Working Time Directive, which was passed against the opposition of the UK government.

THE EUROPEAN COMMISSION

The European Commission consists of one commissioner from each member state who is appointed for four years but is accountable to the European Parliament and not the member state government. The commission can initiate legislation on its own behalf, or when delegated to do so by the Council. Otherwise, its role is to ensure that EU policies are being implemented and advise the Council on what further legislation is required. It acts as a form of civil service and executive, and can take action in the ECJ to ensure that EU laws are complied with.

THE EUROPEAN PARLIAMENT

The European Parliament has 751 members, elected directly from all of the member states. The Parliament has a supervisory role over the Commission and can bring about its resignation on a motion of censure. It has to approve the EU budget and can also ask questions of EU commissioners. As regards legislative power, it is involved in the passing of EU Directives by way of the 'co-decision procedure', now referred to as the ordinary legislative procedure, which is set out below. This is a system that involves working with the Council to come to an agreed position.

THE EUROPEAN COURT OF JUSTICE (ECJ)

The ECJ consists of 28 judges headed by a President. The court is assisted by nine Advocates-General who give legal opinions in order to aid the court. The judicial appointments are for six years, and as with the case of commissioners the judges are independent of any member government control. The court acts as an administrative court, protecting EU citizens against unlawful acts of EU institutions. It also settles disputes between the Commission and member states. Under Article 267 of the Treaty (previously Article 234) the court can give a preliminary ruling on any matter of law raised in a member state court and refer the decision to that member state court to be acted upon.

EU legislation

Although most health and safety legislation is passed through the UK Parliament, in many cases parliament is simply implementing European Law. Since the development of the Single European Market, there has been a requirement for European member states to protect the health and safety of their workers under Article 153 of the Treaty of Rome (previously Article 137). Following the implementation of this article, the EU has played an increasing role in the framing of health and safety legislation for the UK. European laws usually emanate from the Council of Ministers at the behest of the European Commission, before being approved by the European Parliament.

TYPES OF LEGISLATION

The EU introduces law in two main ways. It can still pass EU Regulations, which come from the Council of Ministers and take immediate effect in UK law. Such a process was used to implement the regulation of drivers of goods vehicle and passenger vehicles *(Regulation 3820/85)*. On the other hand, most recent legislation is passed indirectly by the use of EU Directives. This is an instruction to the member states to introduce legislation in their domestic law to implement various broad requirements. A key example here is the *Framework Directive on Health and Safety (Directive 89/391/EEC)* with its relevant daughter directives. This led to the implementation of the so-called 'six pack' of health and safety regulations, namely the regulations covering the management of health and safety, the provision of personal protective equipment, the guarding of machinery, the safety of the workplace, manual handling and the use of display screen equipment.

THE LEGISLATIVE PROCEDURE

The procedure used for the creation of most health and safety Directives is the ordinary procedure, previously known as the co-decision procedure. Under this procedure the Commission first proposes the new legislation and this is passed by the European Parliament for first reading, before being returned to the Council who adopt it by way of a qualified majority vote. It then returns to the Parliament for a second reading at which it can approve the proposal as adopted by the Council, reject it by a majority vote or seek to amend it. If amended, the Council has to accept the amended version by a qualified majority vote. Where there is continuing disagreement between Parliament and the Council, a Conciliation Committee meets to approve a compromise position. If this is obtained, the Directive must then be passed by the Council and Parliament to become law. If there is no agreement then the Directive is not adopted.

PROCEDURE UNDER ARTICLE 114

In addition, some directives are passed under Article 114 of the Treaty of Rome (previously article 95), which was designed to facilitate the approximation of laws in general across the EU. Such directives have to be passed by a unanimous vote of the Council and have been used to pass directives such as the *Directive on Machinery Safety (89/392/EEC)*, which led to the passing of the *Supply of Machinery (Safety) Regulations 1992*.

THE IMPACT OF EU LAW

Directives passed by the EU have had a significant effect on the nature of health and safety law in the UK. Apart from the Framework Directive as mentioned above, there have been a number of key directives passed that have been implemented in UK law. The *Biological Agents Directive (2000/54/EC)* and the *Carcinogens Directive (2004/37/EC)* were implemented through modification of the *Control of Substances Hazardous to Health Regulations 2002 (COSHH)*, while the Physical Agents (Vibration) Directive and the Physical Agents (Noise) Directive were implemented in the *Control of Vibration at Work Regulations 2005 (CVWR)* and the *Control of Noise at Work Regulations 2005 (CNWR)*. The present position is that nearly all UK health and safety law now comes from the European Union, although it is still subject to interpretation in the UK courts. However, it is the case that many of the directives are based on existing UK legislation, as the UK is the pre-eminent country for developing health and safety law. In this way, the *Workplace (Health, Safety and Welfare) Regulations 1992* and the *Provision and Use of Equipment Regulations 1998* effectively reintroduced the old *Factories Act 1961*. So while much of our new law is from the EU, the key elements of health and safety law are still arguably made in the UK.

Case law

The case law here involves both the decisions of the UK courts, as well as decisions of the ECJ. I will look at the two sources of law below and see how they affect the development of health and safety law in the UK.

UK case law

DEFINITION

Apart from legislation, the other key source of law in the UK is case law, sometimes referred to as common law. This derives from the legal principles laid down by judges in key cases. For example, the basic elements of the employer's duty of care for health and safety were established in the leading case of *Wilsons & Clyde Coal* v *English (1937)*. This has now been incorporated into statute as regards criminal liability in Section 2(2) of the *Health and Safety at Work Act 1974*, but civil liability is still mainly based on case law.

TORT OF NEGLIGENCE

Most of the Tort of Negligence is based on legal principles laid down in decided cases. The extension of liability to mental injury is derived from the leading case of *Walker* v *Northumberland County Council (1995)*, while the development of new rules on causation for mesothelioma stems from the case of *Fairchild* v *Glenhaven Funeral Services Ltd (2002)*. Both of these cases illustrate the ability of case law to develop new legal principles and is one of the reasons why the Tort of Negligence has been so flexible, adapting to new concepts of liability as public attitudes shift.

STATUTORY INTERPRETATION

In addition to the developments in the Tort of Negligence, we must also consider the importance of statutory interpretation through case law. Many statutes have terms that are not always clearly defined and it falls to the courts to clarify the situation. Sometimes help can be found in the statute itself, as is the case for Sections 52, 53, 76 and 82 of *HSWA 1974*, but in other cases the courts have to make the decision in the light of current perceptions of fairness.

An example of this is the case of *Close* v *The Steel Company of Wales Ltd (1962)*. Here, Section 14 of *the Factories Act 1961* involved the duty to fence dangerous parts of machinery. However, this was not seen as referring to the prevention of materials being ejected from the equipment, even if they could cause injury to the employee. This meant that there was a gap in the legislative protection, which was not remedied until Regulation 12 of *PUWER* specifically provided for such protection to be afforded.

In more recent times there has been a dispute over the definition of the term 'work equipment'. In *Hammond* v *Commissioner of Police for the Metropolis (2004)*, the term 'work equipment' was confined to those items used by the claimant in his work and so did not apply to the vehicle he was working upon. This decision was eventually overruled in *Spencer-Franks* v *Kellogg Brown & Root (2008)*, which redefined work equipment to include anything used by a person in the workplace, including the self-closing door that the claimant was working on at the time, the failure of which led to his injury.

The definition of terms is very important here as the statute defines the nature of the liability, and under *PUWER* this can often be strict liability rather than fault-based. However, this is now only of importance in defining criminal liability with the ending of the right to sue for breach of statutory duty in respect of health and safety regulation. We can therefore see how statutory interpretation plays a key role in helping to determine the nature of liability in health and safety cases.

European Union case law

Following the implementation of the *European Communities Act 1972*, not only does EU legislation have force of law, but the judgments of the ECJ now have supreme authority in respect of legal matters. As a result, any EU directive has to be implemented in accordance with both the letter and the spirit of EU law. This is because the ECJ adopt a 'purposive' approach to judicial interpretation.

IMPLEMENTATION OF EU DIRECTIVES

As a result of this purposive interpretation, directives must be implemented in UK law to the satisfaction of the EU, or it is possible for the UK to be taken to the ECJ. An example here is the way in which the UK Government was compelled to introduce the *Working Time Directive (93/104/EEC)*, albeit with modifications. The UK government failed in its argument that it was not a health and safety matter and so had to modify UK law in order to comply with the EU directive.

Moreover, if the EU believes that the UK is not fully implementing EU laws it can compel the UK to introduce new legislation to fill the gaps. An example here is the case of safety representatives, where the UK argued that the *Safety Representatives and Safety Committee Regulations 1977* was adequate to comply with the EU requirement for consultation with employees, as laid down in the *Framework Directive on Health and Safety (89/391/EEC)*. The ECJ held that the UK was not complying with the law, because the UK regulations only applied to workplaces with an independent trade union and this excluded the vast majority of workplaces. As a result, the government had to introduce the *Health and Safety (Consultation with Employees) Regulations 1996*, which introduced the concept of the Representative of Employee Safety.

Consequently, the UK has to ensure that it fully implements EU legislation and that such legislation is interpreted in accordance with the principles behind the legislation. It would appear that the UK cannot pass legislation that conflicts with that of the EU; otherwise it is likely to be struck down by the UK courts.

The judicial system in England and Wales (see Figure 1.2)

The judicial system reflects the different types of law, being divided between civil and criminal liability, and the court systems here are also very different as a result. However, we should note that in some cases there is an overlap between the different systems, especially in the case of quasi-judicial sanctions such as enforcement notices. I will also look at the quasi-judicial system itself as it relates to health and safety, which is essentially the Employment Tribunal system. Here I will start by looking at the system as it appertains to England and Wales, before examining the Scottish system. The relevant judicial systems are as set out in Figures 1.2 and 1.3.

The Criminal Law system

Criminal offences fall broadly into three types; summary, indictable and 'either- way' offences. Summary offences are minor offences, which are dealt with by the Magistrates' Court. The main health and safety offences, which are summary only, are such matters as falsely pretending to be an inspector or intentionally obstructing an inspector in the carrying out of his duties under *Section 25* of *HSWA*. Indictable offences include more serious matters such as common law manslaughter and offences under the *CMCHA*, all of which must be heard in the Crown Court. Most offences under *HSWA* are 'either-way' offences, which means that they can be heard in either the Magistrates' Court or the Crown Court. The court to which they will be allocated is actually determined by the Magistrates' Court in Committal Proceedings, on the basis of representations by the prosecution and the defence.

The Magistrates' Court

The Magistrates' Court is a **court of first instance,** which means it deals with cases on the first hearing. Magistrates' Courts in England and Wales are presided over by either a bench of three lay (unpaid) magistrates, or by one stipendiary (paid) magistrate, now known as a District Judge Magistrate.

The role of the Magistrates' Court in respect of health and safety issues is varied. First, in Committal Proceedings it considers the case put before it, to see if there is an arguable case to answer and it then decides which court the case should be allocated to. If it is a summary offence, then it must go to the Magistrates' Court and if it is an indictable offence it must go to the Crown Court. However, if it is an either-way offence, as most offences are under the *HSWA 1974*, then it is up to the Magistrates' Court to either hear the case itself or send it to the Crown Court.

In deciding which court to allocate an either-way offence to, the Magistrates' Court will take into consideration the views of the prosecuting authority and the inspector, as well as those of the defendant. The factors that will determine whether to commit to the Crown Court for trial will include, whether the defendant has committed the same offence before, or where there is evidence that the defendant has deliberately refused to comply with the law. The defendant may seek to obtain a Crown Court trial if they believe they are innocent and want to be heard by a jury. They may also

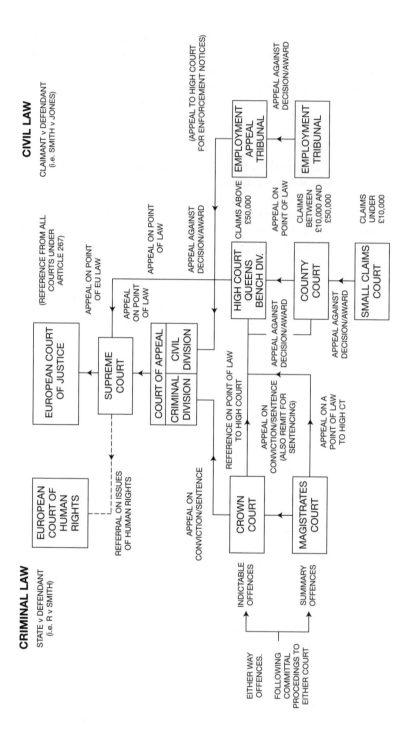

Figure 1.2 The court structure (England and Wales)

Source: Diagram by author.

think they are more likely to be acquitted in the Crown Court, which appears to be borne out by the evidence. This may simply reflect the fact that all the weaker cases are heard in the Magistrates' Court, because the **accused** expects to get a lower sentence if he elects to go to Magistrates' Court rather than Crown Court. Moreover, their legal representative will almost always recommend trial by jury because there is a far greater chance of an acquittal. This is because the magistrates are believed to be 'case-hardened' as they see the same defendants and the same offences time after time, while a jury is far more likely to be influenced by a skilled barrister, especially given that for most of them, it will be their first time on jury service.

The magistrates are responsible for both deciding on the verdict and determining the sentence. Under the *Health and Safety (Offences) Act 2008 (HSOA)* the penalties for most health and safety offences are now a fine of up to £20,000 or imprisonment for up to 6 months. In some summary cases, the penalties are lower, at level 5 on the standard scale. It is also to be noted that magistrates can impose a continuing fine for every day a person continues to commit the offence.

Appeal from the Magistrates' Court against conviction or sentence is to the Crown Court, where it is heard by a judge without a jury. Appeal on a **point of law** is by way of 'case stated', direct to the Queen's Bench Division of the High Court.

The Crown Court

The Crown Court is also a court of first instance, in so far as it hears cases referred to it by the Magistrates' Court. The Court consists of a judge and jury, with the jury deciding on conviction and the judge determining the sentence. While the judge cannot convict the defendant himself, he is able to acquit by withdrawing a case from the jury on the basis that there is no case to answer.

The Crown Court deals with the more serious cases, including all the indictable offences, such as common law manslaughter or Corporate Manslaughter under *CMCHA*. The penalties for breach of *HSWA* and the subordinate legislation is as laid down in *HSOA*. This allows for an unlimited fine and imprisonment for up to two years. Penalties for common law manslaughter are higher and can involve life imprisonment, while penalties under *CMCHA* can involve fines of up to several million pounds. Originally it was hoped to base fines under *CMCHA* on a percentage of business turnover, but this has been abandoned because it was feared that companies could manipulate their turnover figures to reduce their liability. However, even before *CMCHA* was in force, we have seen fines in the millions of pounds, including a £15 million fine for Transco in respect of a gas explosion at Larkhill in Glasgow, which led to four fatalities. Appeal against sentence and conviction in the Crown Court is to the Court of Appeal (Criminal Division), which we will consider later.

The Civil Law system

Civil actions can be divided according to the level of damages claimed. When these do not exceed £50,000 for a personal action, they will usually be allocated to the County Court unless there are special circumstances. Of these, cases involving amounts under £10,000 will be heard through the Small Claims Court, which is a part of the County Court. Higher value claims will go to the Queen's Bench Division of the High Court in respect of actions for Negligence or breach of Contract, which are the main actions in health and safety.

Under the Woolf Reforms,[2] the civil action procedures have been substantially modified to provide a fairer, quicker and less expensive system. The judge now plays a more direct role in managing the progress of the case, by setting strict timetables for action. Under various pre-action protocols, key information must be disclosed by both parties prior to commencing proceedings, with the intention of avoiding litigation if possible. All hearings will fall into one of three types of action: the small claims procedure, which is up to £10,000 for personal injury claims; a fast-track procedure for claims up to £15,000; and a multi-track procedure for claims above this level. In the fast-track procedure the court sets strict timetables for the running of the case, while in the multi-track procedure the court also uses case management conferences to this end. The use of a common expert evidence witness is also required in all of the above.

The County Court

The County Court is the court of first instance for lower-level claims. If the claim is under £10,000, this will be heard in the Small Claims Court by a District Judge in chambers. This is an informal process conducted outside of the courtroom itself. Higher-value cases will be heard in full court, where they will be heard by a County Court Judge or Circuit Judge, who will make the decision as to liability and will award any damages if necessary. Appeal is to the Court of Appeal (Civil) in respect of the decision or the amount of damages awarded, but appeals on a point of law are heard by the High Court.

The High Court

Most personal injury cases involving serious injuries will commence in the Queen's Bench Division of the High Court. As in the County Court, all cases are heard before a judge alone, who decides on both liability and damages. The High Court can also award an **injunction** if necessary, in order to stop a continuing situation, such as the pollution of a river. The High Court also acts as an appeal court on matters of law for both the Magistrates' Court and the County Court.

The quasi-legal system

Employment Tribunals

The role of Employment Tribunals in respect of health and safety is varied. They usually act as a form of civil court dealing with employment issues. As regards health and safety matters, they hear cases brought under the *Employment Rights Act 1996 (ERA)* involving the right of safety representatives not to be dismissed for carrying out their role under Section 100 of *ERA*, or being subject to any detriment for doing the same under Section 44. It is also the tribunal to which an employer is taken if he refuses to make any medical suspension payments.

In addition to this Civil Law role, the Employment Tribunal also serves a Criminal Law role when it acts as an appeal court against the decisions of an inspector to serve an enforcement notice. In this way the tribunal is involved in **quasi-criminal** legal actions and this is because it has an expertise in employment and health and safety matters that is not to be found in the Magistrates' Court, which would normally be expected to take on such a role.

The tribunal itself consists of a legally qualified chairman and two lay members representing respectively employer and employee interests. Despite its composition, the tribunal tends to work very effectively and usually comes to unanimous decisions. However, there have been recent moves to reduce the role of the lay representatives and leave more cases to be dealt with by the chairman alone.

Employment Appeal Tribunal

Appeal from the Employment Tribunal is to the Employment Appeal Tribunal, which deals with appeals on a point of law. This tribunal consists of a High Court Judge as chairman, with two lay members as before. As in the Employment Tribunal, the procedure is still quite informal although more legal representation is creeping in. Appeals against enforcement notices are direct to the High Court in England and to the Court of Session in Scotland.

The judicial system in Scotland

Scotland has its own specific legal system (see Figure 1.3), which eventually links in to the system for the UK as a whole at the upper appellate levels. At the lower level we can see various differences as set out below. Although the *Scotland Act 1998* devolved law-making powers to the Scottish Parliament on a number of domestic matters, health and safety legislation still remains the prerogative of the UK Parliament.

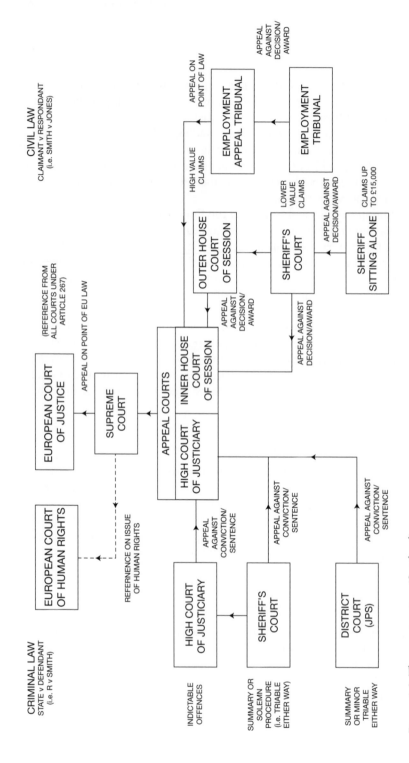

Figure 1.3 The court system (Scotland)

Source: Diagram by author.

Criminal Law system

At first instance, minor offences are heard by the District Court, which is equivalent to the Magistrates' Court in England and Wales. The court is presided over by a bench of Justices of the Peace (i.e. lay magistrates) or a Stipendiary Magistrate. They can impose a fine of up to £2,500 or 60 days imprisonment.

More serious offences are heard at first instance by the Sheriff's Court, this by way of either a summary procedure or solemn procedure. It deals with lesser offences under the summary procedure and can impose a fine of up to £5,000 or three months imprisonment. Alternatively, under the solemn procedure, these courts operate like a Crown Court with a judge (or Sheriff) and jury, and they can pass sentences of up to two years with an unlimited fine. Very serious offences are heard in the High Court of Justiciary on indictment before a judge and jury. The court has unlimited powers of punishment here. Appeals are to the High Court of Justiciary before a panel of three or more judges. This court hears appeals from the lower courts on the grounds of conviction and sentence.

Civil Law system

The Sheriff's Court exercises civil jurisdiction for lower-level claims. It acts as a court of first instance and it also hears appeals from the decision of the Sheriff alone, a type of small claims procedure for amounts of up to £1,500.

The Court of Session acts as a court of first instance for larger claims, with cases heard by the Outer House of the Court of Session. Cases are presided over by a Lord Ordinary. The Inner House of the Court of Session acts as an appeal court, hearing appeals from both the Sheriff's Court and the Outer House of the Court of Session. The Inner House also deals with appeals against enforcement notices, the role carried out by the Employment Tribunals in England and Wales. Appeal from the Court of Session Inner House is to the UK Supreme Court.

The appellate courts

In both Civil and Criminal law there is a right of appeal to the higher or **appellate courts**. They include two UK courts, namely the Court of Appeal and the Supreme Court. In addition, there may be an appeal to the ECJ.

Court of Appeal

This is divided into two divisions, one for Criminal Law cases and the other for Civil Law cases. The Criminal Law Division hears cases from the Crown Court on grounds of conviction or sentence. The Civil Law Division hears appeals from the Queen's Bench Division of the High Court on matters of liability and the amount of any award, as well as some cases direct from the County Court. It also hears appeals from the Employment Appeal Tribunal as regards employment matters. In both courts the decision is made by a panel of between three and five Justices of Appeal with no jury involved.

Supreme Court of Justice

The newly-created Supreme Court replaces the old House of Lords' Judicial Committee and consists of twelve senior Law Lords headed by the President of the Supreme Court. The court hears appeals on a point of law from the Court of Appeal, as well as a limited number of appeals from the High Court, also on points of law.

European Court of Justice

Following the entry of the UK into the European Economic Community (now European Union), the ECJ is now the superior court of law for any matter appertaining to matters of European Law. This means that it will also adjudicate in respect of any UK laws passed to implement EU directives. In the area of health and safety, most of the law is now based on EU Regulations and Directives, most specifically the *European Framework Directive on Health and Safety 89/391/EEC* and the daughter directives passed under it.

In addition, the ECJ adjudicates on Case Law matters. Under Article 267 (previously 234) of the Treaty of Rome, a party to a case in the UK can refer to the ECJ for a 'preliminary ruling' on a matter deemed to lie within the scope of EU law. This must occur once the case has reached the final appeal court in the UK if one of the parties requests this. In the UK, this court is normally the Supreme Court, but it can be the High Court of Justiciary for Scottish criminal cases. The judges in the ECJ will determine how the law should be interpreted in order to comply with EU law, and that decision will be remitted to the UK court. It is then up to the relevant UK court to apply that interpretation when coming to their decision in the case before them.

In addition, all UK laws are subject to scrutiny by the EU and this has led to challenges to existing UK legislation if it does not comply with EU directives. For example, in the case of *Commission* v *United Kingdom (C-127/05)* the ECJ has also adjudicated on whether the qualification of 'reasonably practicable' is consistent with the requirement of EU safety legislation. In this case, the ECJ accepted that the UK could use this qualification, accepting that health and safety law in the EU did not impose no-fault liability.

It can be seen that the ECJ now exercises considerable control over UK health and safety law. It is the superior court in most health and safety matters and its influence is likely to grow in this area as the EU states seek to harmonise their health and safety provision under the aegis of the European Health and Safety Agency based in Bilbao, Spain.

The European Court of Human Rights

The European Court of Human Rights is not part of the European Union and the court has no official position in the UK legal system. However, under the *Human Rights Act 1998* the UK has agreed to abide by the terms of the Convention on Human Rights and to incorporate its principles into any legislation or judicial decisions. This will not normally affect health and safety matters, but the right to a fair trial under Article 6 of the Convention was used as the basis of a challenge to the shifting of the burden of proof in prosecutions under Section 40 of the *HSWA 1974*. In the event, the position of the UK was upheld in the key case of *Davies* v *The Health and Safety Executive (2003)*, as detailed on p. 35.

The nature of liability

In Criminal Law and Civil Law there are various types of liability, which we need to consider here. Some of these are very similar in nature, which can create an element of confusion. The main types of liability are as detailed below.

Strict liability

Some health and safety statutes or regulations impose a form of strict liability. This is where the relevant legislation uses words such as 'shall' or 'will' without any form of qualification, as in Regulation 4 and 5 of *PUWER 1998*. This creates a legal liability which is strict, and liability will attach to the defendant even in the absence of fault. In criminal cases, there is some evidence that the prosecution will fail, unless it is possible to show an element of fault on the part of the accused, as can be seen in the decision in *R* v *Nelson Group Services (Maintenance) Ltd (1999)*, where the employer was able to pass liability onto his employees. In civil cases, the courts are much happier to impose strict liability, as this enables the injured party to more easily obtain redress. However, following the abolition of the right to sue for breach of statutory duty, most of the strict liability in civil actions no longer exists.

Fault-based liability

Fault-based liability is to be found in both Civil and Criminal Law. In the case of Civil Law, the use of the concept of the 'reasonable' man and what is 'reasonable foreseeable', underpins civil liability in the Tort of Negligence. In the case of Criminal Law, the relevant limiting factor is the concept of 'reasonably practicable', which is built into *HSWA* as well as many other safety regulations.

Reasonably foreseeable

In the Tort of Negligence, the liability of the defendant is limited in various ways. First, the duty of care is limited to matters that are 'reasonably foreseeable'. Second, the defendant is only liable for breach of the duty of care if he acts in a way in which a 'reasonable' man would not act. Finally, in determining whether the damage is caused by the breach, once again the court will discount any loss if this is not 'reasonably foreseeable' under the rules of causation. It is the court itself that determines the meaning of the word 'reasonable' here, and this reflects the development of social norms. As a result, the defendant will now be liable for mental injury such as **nervous shock** and mental illness, in a way that would not have been available to the claimant in the past. This was made clear in decisions such as *Walker* v *Northumberland County Council (1995)*, as mentioned on p. 58, where the defendant council was held liable for the claimant's severe nervous breakdown after it failed to deal with his excessive workload.

However, it should also be noted that the courts tend to favour the claimants in cases of personal injury in a number of ways, as set out previously. This will include the requirement for the defendant to take his victims as he finds them, once it is proven that he is responsible for the initial injury. Similarly, as noted above, the courts have been willing to modify the rules of causation, to allow a claim against an employer where there is much less than a 50 per cent chance that he is liable, as in the case of mesothelioma actions such as *Fairchild* v *Glenhaven Funeral Services Ltd (2002)*.

Practicable

Some statutory regulations are qualified by the term 'practicable'. This is a lesser standard than strict liability, but it requires the employer to carry out whatever controls are technically feasible. This was made clear in the case of *Adsett* v *K & L Steelfounders & Engineers Ltd (1953)*, where the employers were held not liable for failing to install a local exhaust system, as they were unaware of how it could be done at the time when the injury occurred. This type of liability can be seen in the requirements of Regulation 11 of *PUWER*. Cost here is not a factor to be considered, although it is questionable whether an employer would be required to implement controls that were totally unreasonable in terms of cost.

Reasonably practicable

Liability under *HSWA*, as well as much of the subordinate legislation, is qualified by the term 'reasonably practicable'. This is a lower standard than the term 'practicable' and involves balancing the risk of the injury that may occur against the cost in terms of money, time and effort to deal with that risk. As such, this introduces a cost-benefit element, as made clear in

Edwards v *the National Coal Board (1949)*. This can be seen in the case of *Associated Dairies* v *Hartley (1979)*, where the Employment Tribunal quashed an improvement notice requiring the defendant to provide safety boots for all his workers, when it was clear that only a few of them worked in an environment where they were likely to suffer a serious foot injury.

It has become increasingly clear that the concept of 'reasonably practicable' is based on what is 'reasonable foreseeable', in so far as the risk itself is limited to those matters that are 'reasonably foreseeable', a point made clear by Lord Dyson in *Baker* v *Quantum Clothing (2011)*.[3] In effect, the risk must be 'material' and cannot include risks that are fanciful, as stated in *R* v *Porter (2008)*. The key difference is that the requirement to provide protection in Criminal Law is limited by the cost of remedial action, while this is not explicit in Civil Law.

We should also note here that the term 'reasonably practicable' is a qualification of the liability of the defendant, and not a form of defence, as was made clear in the case of *R* v *HTM (2006)*. Therefore it is not affected by the requirements of Regulation 21 of the *Management of Health and Safety at Work Regulations 1999 (MHSWR)*, which state that nothing in any statutory provisions should provide a defence to the employer for a contravention of those provisions by way of an act or default of an employee.

Personal liability

Liability under *HSWA* and the subordinate legislation is usually imposed on the employer as an organisation. However, it is possible for the director or another employee to be personally liable. In the case of a director or senior manager, liability exists under Section 37 of *HSWA*, where an offence is committed by a **body corporate**, if this is with the consent, connivance or neglect on the part of any 'director, manager, secretary or any other similar officer of the body corporate'. Neglect here will include 'turning a blind eye' to events that the person knows are happening but does not want to investigate, as this is clearly a dereliction of his duty to manage. An example here is the case of *R* v *P (2007)*, where a senior manager was found guilty when he failed to deal with unsafe fork lift truck operations, on the grounds that he was unaware that they were happening.

In addition to a director or senior manager being personally liable, any employee can be liable under section 7 of *HSWA* if he fails to take 'reasonable care' for the health and safety of himself and other persons who may be affected by his acts or omissions at work. However, in this case, liability is based on Negligence principles and is not qualified by the term 'reasonably practicable'. This means that if Negligence cannot be proved, then there is no liability, as in the case of *R* v *Beckingham (2006)*, where the manageress of a leisure centre was acquitted of liability for a legionella outbreak that caused multiple fatalities.

Corporate liability

Corporations are liable under *HSWA* and the subordinate legislation when their management failures lead to a breach of their legal duties. However, as regards the prosecution of a company for manslaughter, this poses a problem, as there is a requirement that the prosecution attribute the liability to an individual who can be seen as the 'directing mind' of the organisation. This means that a high level of culpability needs to be proved in the case of a named person, equivalent to liability on a personal basis. In a small one-man company this is not normally a problem and it is quite easy to establish that the owner is the 'directing mind'. This can be seen in the successful prosecution of Mr Kite, the director of OLL Limited, in respect of a fatal accident involving the drowning of four sea scouts in Lyme Bay when a canoeing expedition went wrong. However, where the company is very large, it may be very difficult to attribute liability to a specific person. This can be seen in *R v P & O Ferries Ltd (1991)*, the case of the sinking of the Herald of Free Enterprise due to the bow doors being left open. It was found impossible to attribute liability to any one person and so the company itself was not liable for manslaughter in respect of an accident that cost the lives of 189 people.

As mentioned above, the position has now been altered by the passing of the *CMCHA*. This now makes a corporation liable for manslaughter if the way in which it is managed causes a person's death, and this amounts to a gross breach of a relevant duty of care owed by the organisation to the deceased. This means that the corporation will be liable as an entity, even if it is not possible to identify a specific director or manager who can be seen as the 'directing mind' and who would have sufficient culpability to be liable on an individual basis. The new act does not impose individual liability as such, but imposes liability on the organisation, which may be subject to a large fine. The advantage is that the corporation itself is deemed to be liable for manslaughter and so this will be seen as a more serious offence than merely a breach of *HSWA*. So far the act has only been used against small or medium-sized corporations, so we will have to wait to see how it will work when used in the case of a large corporate body.

Vicarious liability

A corporation or employer can be liable for the acts or omissions of its employees, both in Civil Law and in Criminal Law, under the principle of vicarious liability. Nevertheless, the reason for imposing such liability differs between the two.

Civil vicarious liability

In Civil Law, the reason for imposing liability is so that the injured party can obtain compensation for his injury, as the employer must be insured for

the acts of his employees, and so will be in a position
is for this reason that the courts have tended to expand
in this area of law, so that an employer will be liał
employee had disobeyed the express instructions of the
be seen in the case of *Kay* v *ITW (1967)*, when the emp
a fork lift truck driver, who moved a lorry that was bloc
though this had been expressly banned. As a result, the
was behind the vehicle, suffered injury when he was hit b

Criminal vicarious liability

In respect of Criminal Law, the reason for liability is arguably more questionable, especially where an employee deliberately disobeys his employer's
instructions. However, the courts will see this as a failure to provide effective supervision, and criminal liability may still attach, as in the case of
R v *Gateway Foodmarkets (1997)*, where the employer was liable for
the death of an under-manager who had entered a lift housing against the
strict instructions of the employer. As we have noted above, Regulation 21
of *MHSWR* makes it clear that the employer cannot use the default of an
employee as a defence against criminal liability, for breach of any of the
statutory provisions in health and safety legislation.

Joint vicarious liability

Until recently, it was believed that only one party could be vicariously liable
for an employee, as an employer would not be liable for the employees of
subcontractors even though they were under his control. However, this has
now been changed, and the liability may now be shared between contractors
and subcontractors, as made clear in the recent case of *Viasystems (Tyneside)
Ltd* v *Thermal Transfers (2005)*, which I will look at later. This case may
now lead to a need for parties in subcontracting arrangements to clearly
allocate financial responsibility by ensuring adequate insurance provision.

Conclusions

We can see that the legal framework of health and safety law is quite
complex, with an intricate interrelationship between Civil and Criminal Law.
The various types of liability, from strict to fault-based, also create an element
of confusion as Criminal Law definitions may well be carried into Civil Law
actions.

The impact of European Union directives is also of importance, as is the
growing influence of the European Convention on Human Rights with
respect to the presumption of innocence in criminal matters. This means that
health and safety law is increasingly influenced by external factors and legal
principles that may conflict with political imperatives in the UK. For example,

sal of the UK government to fully implement the Working Time ctive can be seen as pitting UK neo-liberal political and economic values gainst the desire of the EU to increase health and safety protection for employees. These conflicting influences will no doubt continue to work themselves out over the next few decades, but we should not forget that UK health and safety law is in itself a major source of, and influence upon, EU health and safety law.

Finally, we should note that the UK Case Law system of judicial law-making will have a key effect in creating the need for the UK government to react to sudden changes in legal liability. The impact of the mesothelioma cases with *Fairchild* v *Glenhaven Funeral Services (2002)* and the contrary decision in *Barker* v *Corus (2006)* were a major factor leading to the government's introduction of the *Compensation Act 2006* to clarify the legal position. This ability of case law to change very quickly will continue to impact on health and safety legislation, making this a very fluid and dynamic area of law.

Notes

1   Professor Ragnar Lofstedt 'Reclaiming health and safety for all: an independent review of health and safety legislation', (November 2011) Cm 8219
2   See 'Access to justice', final report of the Right Honourable the Lord Woolf Master of the Rolls (July 1996)
3   Lord Dyson in *Baker* v *Quantum* (2011) UKSC17 at para 68; and Forlin, G. 'Developments in health and safety' *Archbold Review* 2011, 7, 7–8

2 Criminal liability for health and safety

Introduction

In this chapter, I would like to outline the key criminal liabilities for health and safety. Criminal sanctions for health and safety have a dual purpose: first they are designed to deter employers and employees from engaging in unsafe practices that could lead to injury to themselves or others; second, they are designed to ensure that health and safety is effectively managed so as to prevent such a situation from arising in the first place.

There are two levels of control here. First, there is a general liability for health and safety and the management of such, which is incorporated into the *Health and Safety at Work etc. Act 1974 (HSWA)* as well as the *Management of Health and Safety at Work Regulations 1999 (MHSWR)*. In addition, there is a **result-based** general liability in respect of a death at work, both under gross negligence manslaughter at common law and the new *Corporate Manslaughter and Corporate Homicide Act 2007 (CMCHA)*. The second level of control is the wide range of subordinate legislation passed under the provisions of Section 15 of *HSWA*, which covers specific safety hazards, such as the *Provision and Use of Work Equipment Regulations 1998 (PUWER)* or the *Control of Substances Hazardous to Health Regulations 2002 (COSHH)*. In this chapter I will concentrate on the first type of liability, which is more general in nature, but I will refer to some of the common aspects of the more specific regulations. This subordinate legislation is covered in more detail in Chapter 5.

The Health and Safety at Work Act etc. 1974 (HSWA)

Scope of the legislation

The *HSWA* was the product of the Robens Report of 1972,[1] which looked at the state of health and safety legislation in the United Kingdom at the time. The report concluded that the existing system was both chaotic and excessive, with too many regulations covering different types of workplace, sometimes conflicting in nature. In addition, the law was too prescriptive, concentrating on a set of specific rules enforced by the inspecting body.

The new *HSWA* was designed to cover all persons at work (with a few exceptions such as domestic servants), including both employees and the self employed, in total some 15 million employees at the time. It imposed duties not only on employers, but also on manufacturers and suppliers of goods for use at work, as well as employees themselves and even members of the public, as in Section 8, which prohibits interference with anything provided in the interests of health and safety. Instead of relying on prescriptive rules, the legislation was 'goal setting' in nature, laying down general requirements to ensure the health and safety of employees and third parties, but leaving it up to employers and other parties to determine how to meet these requirements.

The original concept was that there would be simply one piece of legislation, but this would be supported by a series of Codes of Practice, which would help to explain how the law could be complied with for specific areas, such as work equipment and personal protective equipment. However, when the UK joined the European Community, the preference of the other member states was for more prescriptive, legally enforceable regulations rather than non-legal codes, so the UK was forced to introduce a raft of supplementary legislation. Originally, the penalties for breach of health and safety legislation were mainly financial, as imprisonment was only possible for a few offences, such as failing to comply with an improvement or prohibition notice. However, this is no longer the case following the implementation of the *Health and Safety (Offences) Act 2008 (HSOA)*, which provides for imprisonment for up to six months for summary conviction and two years on indictment in respect of breaches of Sections 2 to 8 of the act.

The scope of the liability covers that due by the employer to employees, by the employer to non-employees, by the occupier of work premises to all parties, by a manufacturer, importer or supplier of articles and substances for use at work to all parties and by an employee to himself and others. This can be seen in the different sections of the acts as set out below. Criminal liability under *HSWA* is not strict, but is qualified by the term 'so far as is reasonably practicable' in respect of Sections 2, 3, 4 and 6 of the act. In Section 7, the liability is similar to that of common law Negligence and based on taking reasonable care, while Section 8 is based on the normal criminal standard of intent or recklessness.

The burden of proof

The burden of proof is on the prosecution to prove its case, and the standard of proof required in general is **'beyond all reasonable doubt'**, reflecting the fact that a criminal conviction is a serious issue and reflects an element of 'moral culpability'. However, we should note that under Section 40 of the

HSWA the burden of proof may be switched to the defendant once a '**prima facie**' breach of the *HSWA* is proved, requiring the defendant to prove that he has done everything 'reasonably practicable' to avoid liability. However, once this burden is shifted to the accused, it can be discharged on the basis of the civil standard of the '**balance of probabilities**'. Indeed this situation applies for all legislation, where liability is qualified by the term 'reasonably practicable'.

This transfer of the burden of proof from the prosecution to the defence was challenged in *Davies* v *The Health and Safety Executive (2003)*, when it was argued that this was in breach of the European Convention of Human Rights, because in criminal law, the normal position was that the defendant was presumed to be innocent before being proved guilty. However, the Court of Appeal held that the switching of the burden of proof was not in breach of the Convention, and that it was justified and necessary given the regulatory nature of the legislation. However, this decision was made when the *HSWA* did not usually impose custodial sentences. With the introduction of the *HSOA*, the range of custodial offences has been considerably widened and it is unclear whether the court would come to the same decision today.

Duties of the employer to employees

Section 2

Section 2(1)

The general duty of care is set out in Section 2(1), which states that 'it shall be the duty of every employer to ensure, so far as is reasonably practicable, the health, safety and welfare at work of his employees'. This is a comprehensive duty and should be seen as the criminal equivalent to the common law duty of care in Negligence, except that the liability is qualified by what is 'reasonably practicable', not by what is 'reasonably foreseeable'.

The definition of the term 'employee' has been extended to include government trainees and persons on work experience placements by virtue of the *Health and Safety (Training for Employment) Regulation 1990*. Later on it was extended to Police Officers under the *Police (Health and Safety) Act 1997*. In addition, the act can apply to activities outside of Great Britain, as laid down in the *Health and Safety (Application outside Great Britain) Order 2001*.

As regards the definition of the term 'at work', it would appear that this does not cover a person on the way to work, even if they are actually on the employer's premises, so long as they have not 'clocked on', as can be seen in the case of *Coult* v *Szouba (1982)* discussed on p. 43. It should be noted that a person working on an oil rig will be deemed to be 'at work', for the purposes of giving information on health and safety by virtue of the *Offshore Installation and Pipeline Works (First Aid) Regulations 1989*.

Finally, it is important to note that an employer will still be liable under the act even where there are no employees actually at work at the time of the breach, as the offence is of a continuing nature. This was established in the case of *Bolton M.B.C.* v *Malrod (1993)*, where the employer was prosecuted in respect of a faulty asbestos decontamination unit, following a visit of the HSE, even though work had not commenced at the time.

Section 2(2)

The general duty in Section 2(1) is now further split into more particular duties under Section 2(2). The division here is based on, but is not the same as, the categories set out for the common law duty of care in *Wilsons & Clyde Coal Ltd* v *English (1938)*. These are all qualified by the term 'so far as is reasonably practicable', and they are shown in the following sections.

SECTION 2(2) (A) SAFE PLANT AND SYSTEMS OF WORK

Under Section 2(2)(a) the employer is required to ensure 'the provision and maintenance of plant and systems of work that are . . . safe and without risks to health'.

The term 'plant' here is defined in Section 53(1) of the act as 'any machinery, equipment or appliances'. These would appear to include machinery used in the work activity, but not machinery that is being worked upon as this would be work equipment, as established in the case of *Haigh* v *Charles Ireland Ltd (1974)*, where a safe being cut open by a scrap merchant was not deemed to be work equipment.

The definition of work equipment is quite wide and will include a delivery vehicle as in *Bradford* v *Robinson Rentals (1967)*. More recently it has been established that the definition of work equipment will include almost anything that is used by the employee in the course of his employment, such as, for example, a self-closing door as in *Spencer-Franks* v *Kellogg, Brown & Root (2008)*. It should also be noted that an employer is liable under this section, where he fails to supply any equipment. In addition, under the *PUWER*, there is a requirement that the work equipment be maintained in a safe working condition, and this is implied into Section 2(2)(a) of *HSWA*.

As regards a safe system of work, this involves a formalised system of working procedures that will avoid reasonably foreseeable hazards. In this way, a practice such as filling flammable gas canisters in close proximity to a source of ignition would be seen as a failure to provide a safe system of work as per *R* v *Portagas (1984)*. However, it should be noted that a system of work is not unsafe simply because there are still risks involved, as this is intrinsic in certain activities. For example, the job of feeding the tigers at a zoo always carries the potential of a fatal attack on the keeper, but this in itself would not mean that a system of work for this activity was in itself unsafe, as made clear in the case of *Langridge* v *Howletts Zoo & Port Lympne Estates Ltd (1996)*.

SECTION 2(2)(B) HANDLING, STORAGE AND TRANSPORT OF ARTICLES AND
SUBSTANCES

Section 2(2)(b) requires the employer to ensure the 'safety and absence of risk to health in connection with the use, handling, storage and transport of articles and substances'.

In Section 53(1), articles are defined as items of plant and component parts, while substances include natural or artificial substances, whether in the form of a solid, liquid or vapour. Much of this area of law is now further developed by the *COSHH* or the *Classification, Labelling and Packaging Regulations 2010*, especially in respect of transport of hazardous chemicals. As such, the legislation involves the introduction of safe storage systems and means of transportation, as well as safe systems of work relating to the carrying and handling of such items.

SECTION 2(2)(C) PROVISION OF INFORMATION, INSTRUCTION, TRAINING AND
SUPERVISION

Section 2(2)(c) imposes a wide ranging duty involving three interrelated aspects, namely the provision of information, training and adequate supervision. The requirement for the provision of information is also linked to consultation and is further developed in the *Health and Safety (Consultation with Employees) Regulations 1996*, which require the employer to consult employees about any measure that would substantially affect their health and safety. The provision of information is also covered in more detail in Regulations 10 and 12 of *MHSWR*. Note that the information given must be meaningful and accurate otherwise liability will be incurred, as in the civil case of *Vacwell Engineering* v *BDH Chemicals (1971)*, where the defendants were held liable for an explosion resulting from inadequate warnings on a consignment of chemicals.

The need for training is a natural development of the provision of information, and such training should focus on specific task-related information. Again this is developed in more detail in Regulation 13 of *MHSWR* as well as in other specific supplementary regulations such as Section 12 of the *COSHH* or Regulation 9 of *PUWER*. Failure to provide adequate training will be seen as a breach of Section 2(2)(c), as made clear in the case of *R* v *Swan Hunters Ltd (1982)*.

The issue of supervision is a slightly different matter, as it is a wider management function and linked to the element of control as required in proprietary management systems such as HSG65. It is also implicit in the general duty of care under section 2(1) of *HSWA*.

SECTION 2(2)(D) PROVISION OF A SAFE PLACE OF WORK

Under Section 2(2)(d), in respect of any place of work, under the employer's control, there is a requirement to ensure 'the maintenance of it in a condition that is safe and without risks to health and the provision and maintenance of means of access to and egress from it that are safe and without such risks'.

This is a duty of care that only relates to premises under the employer's control, otherwise the duty will lie on the controller of the premise under Section 4 of *HSWA*. The duty applies not only to the place of work, but also access to and egress from that place of work, so that it will include the access route to the workplace and means of escape in the case of a fire or other emergency. The duty is supplemented by the requirements of the *Workplace (Health, Safety and Welfare) Regulations 1992 (WHSWR)*.

SECTION 2(2)(E) MAINTENANCE OF A SAFE WORKING ENVIRONMENT

Section 2(2)(e) requires the employer to ensure 'the provision and main-tenance of a working environment for his employees that is . . . safe, without risks to health and adequate as regards facilities and arrangements for their welfare at work'.

This section goes beyond the original scope of the old *Factories Acts* and encompasses two key areas into which health and safety has extended into in recent times: namely environmental and welfare matters. As regards environmental issues, this involves matters such as ensuring the workplace is protected against excess noise, noxious fumes and dusts, as well as ensuring the provision of adequate lighting, heating and ventilation. Again, a variety of subordinate legislation also covers these issues including the *WHSWR* as mentioned above. The issue of welfare facilities covers such matters as washing facilities, sanitary facilities and changing rooms, as well as rest rooms for pregnant mothers. Again the *WHSWR* covers much of this area in more depth.

Section 2(3) safety policy

In addition to the general safety requirements, there is also a duty to provide a safety policy, which has to be written if there are five or more employees. The case of *Osborne* v *Bill Taylor of Huyton Ltd (1982)* clarifies this point, by making it clear that where there are a number of premises involved, but they are all controlled by one organisation, then the number of employees will be the total of employees at all of the premises. The requirement for a safety policy leads on to the need for an effective safety management system, which is expanded upon by Regulations 5 to 12 of the *MHSWR 1999*.

Section 2(6) and 2(7) safety representatives and safety committees

These are duties imposed on the employer in respect of safety representatives as elected by recognised Trade Unions. These requirements are supplemented by the provisions of the *Safety Representatives and Safety Committee Regulations 1977*, which provide for the election of such representatives. Under Section 2(6), employers are required to consult such representatives and under Section 2(7) they have to set up a safety committee if requested to do so by two such safety representatives.

These regulations have been further supplemented by the *Health and Safety (Consultation with Employees) Regulations 1996*, which provide for the recognition of Representatives of Employee Safety, elected by the workplace, where there is no independent Trade Union. Such representatives have similar rights to time off for training, but do not have the right to carry out inspections or to call for the setting up of a safety committee, although this may be done if the management wish to do so. The role of the safety representatives is dealt with in more detail in Chapter 6, which deals with enforcement of criminal liability.

Duty of the employer to non employees

Section 3

In addition to the duty of the employer to his employees, there is a duty on the employer in respect of non employees, which is very wide in scope. This is essentially contained in Section 3(1) of *HSWA*.

Section 3(1) duty of employer to non employees

Under Section 3(1), the employer is under a duty 'to conduct his undertaking in such a way as to ensure . . . that persons not in his employment who may be affected thereby, are not thereby exposed to risks to their health and safety'.

This section has been very widely interpreted by the courts, in such a way as to impose a considerable liability on the employer. Not only is he responsible for the actions of his own employees, he is effectively responsible for the actions of the employees of third parties, such as a contractor working in his undertaking.[2] In addition, an employer will be liable not only for persons at work, but also lawful visitors who use the premises, as was found in the case of *R* v *Trustees of the Science Museum (1993)* (see p. 42).

Such liability will be imposed where the employer is seen as failing to supervise the actions of third parties, such as where he fails to ensure that they comply with the requirements of a permit to work. This was the case in *R* v *Associate Octel (1995)*, where an employee of a contractor was seriously injured, after using a piece of equipment on site that should have been prohibited under the permit. As a result, the site operator was liable under Section 3(1) of the act. Similarly, the employer will be liable where his employee fails to supervise the actions of a contractor, even where the contractor has specialist knowledge. This was the case in *R v British Steel (1995)*, where the shift manager of the defendant employer failed to effectively supervise the operations of a contractor who was erecting a steel gantry, with the result that it collapsed, causing the death of one of the contractor's employees. In this case British Steel was prosecuted for a breach of Section 3(1) of *HSWA*, due to the failings of its shift manager.

When looking at Section 3(1), the court will interpret the liability to include all the particular duties covered in Section 2(2), such as provision of safe plant and equipment and safe systems of work. The focus here is on the conduct of the employer's undertaking and so this will include any activity he engages in, whether on or off his site. It would appear that liability for a **third party** will only be excluded where the evidence is that the employer could not have been expected to have supervised their activity, because the employer did not have the required specialist knowledge. In this way the criminal liability reflects the civil law position, where there is no liability for the faults of specialist contractors. This was demonstrated in the civil law case of *Haseldine v Daws (1941)*, where the occupier of premises was held not liable for the failure of a lift that had just been serviced by specialist lift repairers.

On the other hand, it should be noted that an employer may have a wider liability to members of the public than to employees or contractors, as such persons may have little knowledge of the likely hazards in the workplace. In *R v B&Q (2005)* the court held that this was the case in respect of liability to a lawful visitor to the premises, who was injured by a fork lift truck, whereas the court held that the employer would not have been liable to an employee in the same circumstances.

Section 3(2) duty of self employed person to himself and others

Under Section 3(2), a self employed person has a duty to 'conduct his undertaking in such a way as to ensure ... that he and other persons (not being his employees) ... are not thereby exposed to risks to their health and safety'.

In the case of the risk to themselves, the HSE has long been reluctant to take any action and the Lofstedt Report specifically recommended that self employed persons posing no risk to others should not be liable.[3] This is now implemented in law by virtue of the *Enterprise and Regulatory Reform Act 2013*. However, in the case of a risk posed to other parties, the need for a statutory duty of care is quite clear. This mirrors the position in civil law where a self employed person would be expected to compensate a third party injured by his actions and so is required to take out public liability insurance to cover such an eventuality.

Duties of occupiers of premises

Section 4

In Section 4, the law imposes a duty of care on the controllers of non-domestic premises in respect of non employees who use the premises as a place of work, or who use plant and equipment provided there. The duty is to ensure that the premises are safe, including access to and egress from those premises, so far as is reasonably practicable.

The duty is the criminal equivalent of the civil liability under the *Occupiers Liability Act 1957* and is based on control over the premises. Under Section 4(2) of the act, 'control' involves the obligation to maintain or repair the premises or to ensure the safety of plant or substances in the premises. The term 'premises' includes not only land and buildings, but also vehicles, vessels and aircraft, as well as offshore installations. It will also include communal areas in a block of flats, as decided in the case of *Select Management* v *Westminster City Council (1985)*. The duty includes access to and egress from the premises, so staircases and access platforms will be included.

The liability in respect of any plant or equipment used is owed to third parties, so could include visitors to the premises, such as students using equipment at a college. However, no liability will ensue where the occupier has given clear instructions as to how the equipment should be used. We can note here the case of *Austin Rover Group* v *Inspector of Factories (1989)*, where the defendant had employed contractors to clean a spray booth on his premises, but the contractors had disobeyed the instructions and one of their employees had been killed. The prosecution against the occupier was quashed, a case that can be compared with *R* v *Associated Octel* above.

Liability on designers, manufacturers, importers and suppliers

Section 6

Section 6 imposes a duty of care on a wide variety of parties, including designers, manufacturers, importers and suppliers in respect of articles or substances provided for use at work. In many ways this reflects the liability of such parties to non trade customers under the *Consumer Protection Act 1987*.

As regards articles, the duty of care is defined in Section 6(1) of the act and involves ensuring that such articles are safe and without risk to health. Also there is a requirement to carry out such testing and examination as is necessary to ensure the same and to provide adequate information about their use. This would include the provision of information on the safe use of equipment, such as the welding equipment leased out to a contractor by Swan Hunters Shipbuilders in the case of *R* v *Swan Hunters Ltd (1981)*.

In addition, under Section 6(2), the designer or manufacturer of such articles has a wider duty to ensure that any necessary research is carried out to eliminate or minimise any risks to health posed by such articles. As such, the law imposes a continuing duty on the designer and manufacturer to improve their product over time.

Under Section 6(3) there is a duty of care on the erector or installer of an article for use at work to ensure that nothing in the way in which it is erected makes it unsafe or a risk to health. This is a section that will apply to parties erecting scaffolding within a workplace, or even the operator of a fairground ride.

In respect of the use of substances at work, Section 6(4) imposes a duty to ensure that a substance is safe and without risk to health, by carrying out the same testing and examination as applies in the case of articles. This will also involve the provision of adequate information, which has to be clear and not misleading. This can be a problem in the case of biological substances such as pathogens where the risk may be difficult to quantify. Nevertheless, the defendant will have to take these issues into consideration, as made clear in the case of R v *The Trustees of the Science Museum (1993)*. In this case the defendants were liable for the failure to maintain their air conditioning system, with the result that the public was exposed to the possibility of contracting legionnaires' disease. As with articles for use at work, there is a continuing obligation to carry out research as set out in Section 6(5) of the act.

The term 'supplier' under the act is defined in Section 6(9) as being the 'effective' supplier and not necessarily the ostensible supplier in law. This means that it excludes financial organisations that may be deemed a supplier in law, but are really only providing credit facilities through a hire purchase arrangement. Similarly, under the *Health and Safety (Leasing Arrangements) Regulations 1992* a leaseback arrangement is also covered, so as to ensure that the ostensible supplier in law is not affected.

Duty of employees

Section 7

In addition to the duties imposed on the employer, manufacturer and controller of premises, the *HSWA* also imposes key liabilities on employees on a personal basis. This is a departure from the previous arrangements under the *Factories Acts*, where the liability was only on the employer. It is important to note that in the case of the employee, the liability is not qualified by the term 'so far as is reasonably practicable'. Instead the duty placed on the employee in Section 7(1) is to take reasonable care for himself and others who might be affected by his acts or omissions, in effect a criminal version of the common law duty of care in Negligence. In addition, in Section 7(2), there is a duty to co-operate with the employer in respect of any duty imposed on the latter as regards health and safety.

The duty applies only while the employee is 'at work', so it will not usually apply when the employee is on his way to and from work, even if he is actually on work premises. Therefore, in the case of *Coult v Szouba (1982)*, the employee was not in breach of Section 7 when he was found to be driving dangerously along an internal road on the employer's steelwork plant, as he had not yet clocked in to work. Failure to comply with Section 7 means that the employee is not only liable to prosecution, but is also in breach of his contract of employment and can be dismissed, as this would be seen as automatically fair dismissal under the *Employment Rights Act 1996 (ERA)*.

A prosecution under Section 7 can be taken against employees at any level in the organisation, including senior managers. This can be seen in the case of *R v Beckingham (2006)*, where the manageress of a leisure centre was prosecuted following a legionella outbreak that caused several fatalities. The standard of care here is that of civil law Negligence and so is similar to common law gross negligence manslaughter.

The liability of employees has been further extended under Regulation 14 of *MHSWR*, which includes a positive duty on employees to report health and safety hazards to the employer. This may be implied from the requirements of Section 7(2) of *HSWA* but is not made explicit, so in that respect the duty on employees is slightly wider under the regulations.

Duties on directors and senior managers

Section 37

If an offence is committed by a corporate body, and this is proved to be committed with the consent, connivance or neglect of a director or senior officer of that corporation, then that person may be liable under Section 37 of *HSWA*. In the case of *Armour v Skeen (1977)*, for example, the Director of Roads for Strathclyde Regional Council was held liable under Section 37 for the failure to properly manage a construction project, which led to the death of a worker who fell from a road bridge.

The liability of directors under Section 37 has been much discussed following the introduction of the *CMCHA*. As there is no individual liability under this new act, the temptation is for directors to be prosecuted individually, either for gross negligence manslaughter or under Section 37 of *HSWA* following a fatality at work. Liability under either of these areas of law may lead to imprisonment for the director, so there is a tendency for the defendant to try and pass liability onto the corporation, where the penalty is purely financial. This can lead to a conflict of interest between the corporation as a defendant and any directors, a situation which came to the fore in the prosecution of Lion Steel Equipment, following a fatal fall.[4]

The terms consent or connivance suggest an active role in allowing or encouraging the action that led to the breach, but this is not clear in the term 'neglect'. However, the liability of the director does not require active participation in the commission of the offence, but will include an attempt to avoid responsibility by failing to take action. The term 'neglect' here, involves not only the turning of a 'blind eye' to what is happening, but also failing to be aware of a problem that a director should reasonably be expected to be aware of. In this respect the standard of care is as set down in Negligence and will depend on the level of expertise a director purports to have. So in *R v P (2007)*, a director was prosecuted for failing to stop children being carried on fork lift trucks at the docks and for allowing such trucks to carry unsecured loads, the result of which was a fatal accident to a child. The court felt that any reasonable director should have been aware of what was going on in the workplace, and that such a director would know that these activities were unsafe.

However, the court will not impose liability on a person who does not have any real power to make executive decisions. In *R v Boal (1992)*, the assistant manager was left in charge of a bookshop while the manager was on holiday. The premises were visited by the fire brigade who found several breaches of the *Fire Precautions Act 1971*, as a result of which they prosecuted the assistant manager under Section 37. The prosecution was quashed on the grounds that the assistant manager had no responsibility for fire safety issues, which were always dealt with by the manager. The key issue here is control; the more real control that the director has over events, the more likely it is that the director will face prosecution under this section.

Duties of other parties

Section 36

In addition to a director, a third party can also be prosecuted if he is responsible for the commission of an offence by another party, such as the employer. This allows for the prosecution of a safety manager in the organisation, or more particularly, a safety consultant engaged by the organisation on a self employed basis. An example of this is the 'Fatty Arbuckles' case where a safety advisor was prosecuted when he failed to pick up on serious electrical faults at a client's premises, which led to a fatality.[5] The provision of poor advice can therefore render a safety consultant or advisor personally liable under this section of the act.

Section 8

Section 8 of *HSWA* imposes a duty on all persons, whether at work or otherwise, 'not to intentionally or recklessly interfere with or misuse anything provided in the interests of health, safety or welfare'.

This liability will extend to members of the public who interfere with safety devices such as fire extinguishers or fire escapes. The duty here is not qualified by the term, 'reasonably practicable', but instead is based on the criminal standard of liability, being intent or recklessness. It should be noted that recklessness does not require a deliberate act but can include a failure to use known safety procedures in the case of an experienced worker, as in *Ginty* v *Belmont Building Supplies Ltd (1959)*.

Vicarious liability of the employer

The fact that an employee or safety consultant is seen as being in breach of the law does not mean that the employer totally escapes liability. An employer will be vicariously liable for the actions of his employees and other persons under his control, because the duty of care under Section 2 of the act is non delegable. This is now clear from Section 21 of the *Management Regulations* in respect of any relevant statutory provisions.

As a result, the employer will still be liable, even when an employee deliberately disobeys the instructions of the employer. This can be seen in the case of *R* v *Gateway Foodmarkets Ltd (1997)*, where the employers were liable for a breach of Section 2 of *HSWA*, in respect of the death of an under manager who fell down an open lift shaft while entering the area, in contravention of clear rules to the contrary. The court felt that the employer was aware that the rules were being broken and had chosen to do nothing about it.

Clearly there is a danger that an employer will try to evade liability by blaming the employee for any breach of the law. To allow the employer to evade liability in this way would seriously undermine the effectiveness of the legislation. However, it is clear that the employer can use the failings of the employee as a defence, if he can show that as an employer he did everything 'reasonably practicable' to ensure the health and safety of his employees. This can be seen in the case of *R* v *HTM (2006)*, where a roadway repair organisation were held not to be liable for the death of two of their employees who were electrocuted while moving a mobile lighting tower. It was clear to the court that these were experienced employees, who had acted in breach of clear instructions as to how the tower should be moved, and the evidence was that the employers provided effective training and supervision. The court held that they should not be held responsible for the unexpected actions of trained and experienced employees.

While it is usually seen as important in Criminal Law not to impose liability on the employer in the absence of fault, it does appear that the occurrence of an accident will put the employer in a very difficult position. We have noted that the operation on Section 40 of *HSWA* switches the burden of proof onto the defendant, once a prima facie breach of the law is proved by the prosecution. The employer will then have to prove that it was not 'reasonably practicable' to avoid the breach of the law in order to provide a defence. Therefore in the case of *R v Chargot (2008)*, the employer was liable for a breach of Section 2 of *HSWA*, following a fatal accident to a dumper truck driver, even though it was unclear what had caused the accident. However, the fact that the driver was not wearing a seatbelt at the time of the accident, and evidence that the employer provided inadequate training, were sufficient to render the employer liable. This case has been seen as imposing a very high level of liability on employers similar to strict liability.[6]

Manslaughter

Manslaughter now involves two types of liability in Criminal Law in respect of health and safety. First, there is the liability on an individual, and to this has recently been added a corporate liability as outlined below.

Individual liability

Nature of the liability

In addition to the liability of employees under *HSWA*, there has long been a criminal liability on all persons under common law for gross negligence manslaughter. This liability will apply where a person causes the death of another by gross negligence or recklessness, intention being effectively a different matter and outside of the scope of health and safety law. It is of note here that liability is effectively result-based and only applies where a fatality occurs. There is no criminal liability outside of *HSWA* in respect of negligent actions that lead to serious injury, although some writers have argued that such a liability is overdue.[7] Indeed there is a case for making an employee liable for Grievous Bodily Harm or causing Actual Bodily Harm if their actions or inactions lead to such a result. However, at present the only liability here remains under *HSWA*.

The requirements for a conviction for gross negligence manslaughter were laid down by Lord Mackay in the leading case of *R v Adomako (1994)*. The prosecution has to show that the defendant owed a duty of care to the deceased, that the defendant breached that duty, that the breach caused the death and that the nature of the breach revealed a level of negligence so gross that it deserved to be the subject of criminal liability.

It can be seen that the first three elements of the definition are based on the requirements for proving civil liability in common law Negligence. The duty here is personal, so that there is no vicarious liability for the failings of other parties, unlike the liability under Section 37 of *HSWA*. The key issue here is whether the defendant can be said to have assumed responsibility for the health and safety of the deceased. In the case of a director or senior manager in an organisation, it may be difficult to deny that they have direct responsibility for the actions of the employees involved. However, in the case of an ordinary employee the position is more uncertain, although there are situations that clearly would apply. We may argue here that breach of his liability under Section 7(1) of *HSWA* would be evidence of a breach of a duty of care to the deceased. Similarly, a breach of Regulation 14 of the *Management Regulations*, in respect of his failure to report a serious and obvious hazard to the employer, which then led to the death of a person, may also be seen as constituting a breach of his duty of care, sufficient to **warrant** criminal liability.

The standard of care

In determining whether there is a breach of duty, the standard of care will be that expected of a person in that position in the organisation. In the case of a director or senior manager, the standard will be higher than that of an employee. The standard of care here will be based on the one set out by the HSE and the Institute of Directors in their Guidance for Directors.[8] The factors determining whether there is a breach will include the proximity of the defendant to the deceased and how foreseeable was the accident, these being issues also contributing to the definition of the scope of the duty of care itself. In determining causation, the court will look at whether the breach by the defendant was a 'material factor' contributing to the accident, as made clear in *R v Porter (2008)*. All of the above factors are the same as those in determining civil liability in common law Negligence.

Finally, in determining whether the breach is sufficiently 'gross' as to merit criminal liability, the court will consider whether the failings of the defendant are of such a serious nature as to merit prosecution. In *R v Kite & OLL Ltd (1994)* the defendant had caused the death of four sea scouts in Lyme Bay while on a canoeing trip, as a result of failing to take adequate safety precautions, such as informing the coast guard of where he was going. The court felt that the defendant had organised his operations so badly, that any reasonable person should have been aware of the danger, so this merited criminal liability. It would appear that liability will often apply where the defendant has a duty of care to particularly vulnerable parties, such as young children or trainee employees.

In many cases, the HSE prefer to bring a prosecution under Section 7 or 37 of *HSWA*, as it will usually be easier to prove liability. Indeed, the HSE have sometimes failed to take criminal action for manslaughter, even in the most blatant cases, because they think it will be too difficult to prove. A classic example is the case involving Simon Jones, the student who was killed on his first day at work on the defendant's wharf, having failed to be provided with any real training in how to do a very dangerous job.[9] The HSE initially refused to prosecute the defendant general manager on the grounds that they felt they could not prove their case. In the event, they revised their decision following a well publicised campaign and a **judicial review** of their action. Nevertheless, the jury still acquitted the manager at the re-trial, although the company was found liable for breaches of *HSWA*.

Corporate liability at common law

Apart from the individual liability, a corporation may be liable for gross negligence manslaughter under common law. However, the difficulties in bringing a prosecution for Corporate Manslaughter have been much greater.[10] In order to find a corporation liable for manslaughter, it is necessary to prove the liability of a person in the company who is deemed to be the 'directing mind' of the organisation; in other words, a person who has direct control over operations and who makes the key decisions. However, in a large corporation power is usually dispersed between a number of directors and senior managers, and it is very difficult to establish a single person who can be designated the 'directing mind'. As a result, between 1969 and 1992 there were over 18,000 fatalities at work and yet not a single successful prosecution was brought for Corporate Manslaughter, while there have only been six prosecutions in the last ten years, nearly all for small companies.[11]

The problem is complicated by the fact that in UK law it is not possible to 'aggregate' the liability of a number of individuals in an organisation to attach liability to the organisation as a whole; instead one individual has to be liable on an individual basis to make the corporation liable. While this might be possible in a small 'one-man-band' company, it was usually not possible in the case of a large public limited corporation with a board of directors.

This situation can be seen in the case of *R v P & O Ferries Ltd (1991)*, which was the unsuccessful attempt to prosecute P&O Ferries for Corporate Manslaughter in respect of the sinking of the Herald of Free Enterprise, an accident which led to the loss of 189 lives. The accident was caused by a number of factors, including a deliberate policy of setting sail with the bow doors open in order to allow a quick turnaround, as well as the failure to install a warning system at the bridge to indicate that the bow doors were still open. These were decisions made by the board of directors as a whole and not one individual as such. The whole safety management system was quite inadequate, as made clear by Lord Justice Sheen, who described the safety arrangements as infected by 'the disease of sloppiness'.[12] However, prosecutions could only be brought against the captain and the bosun, who alone had sufficient individual responsibility to merit prosecution. As none of the directors was personally liable, it was not possible to bring an action for Corporate Manslaughter against the company itself. In the event, the manslaughter charges were dropped against the captain and the bosun, as it was felt that the persons actually responsible were avoiding liability, and it would be wrong to put all the blame on individuals much further down the chain of command.

The problems in the P&O case were repeated in the case of *R v Balfour Beatty Rail Infrastructure Services Ltd (2006)*, in respect of the Hatfield rail crash. Once again, the fact that it was not possible to bring charges against a specific director of the company meant that it was not possible to convict the company of Corporate Manslaughter, even though this was seen by the trial judge Mackay J as 'one of the worst examples of sustained, industrial negligence in a high risk industry'.[13] It was clear that the existing law was inadequate, so new legislation was introduced to remedy the situation.

Corporate Manslaughter and Corporate Homicide Act 2007 (CMCHA)

The *CMCHA* seeks to deal with some of the problems caused by the deficiencies in the common law action for Corporate Manslaughter. Most importantly, it focuses on the failure of safety management, rather than the failure of an individual, and in this respect it is very similar to the principles underlying *HSWA*. Indeed there is no individual liability under the act, thus a separate manslaughter prosecution would have to be taken if it was felt a single individual merited prosecution.

Under Section 1(2) of *CMCHA*, the act applies to an 'organisation' but the definition of 'organisation' is drawn very widely, and includes not just corporations as normally understood in law, but also government departments, the police force, partnerships, trade unions and employer organisations.

An organisation is liable under Section 1(1) of the act if the way in which its activities are managed or organised causes a person's death and is also 'a gross breach of a relevant duty of care owed by the organisation to the deceased'. However, in limiting liability to the organisation and its management, the act focuses on the actions of senior managers. There is no vicarious liability here for the actions of junior staff as in *HSWA*, as they are not seen as being responsible for overall safety management.

The duty of care is defined in Section 2 as a duty owed under the law of Negligence in respect of various situations. These include: a duty owed by an employer towards an employee or a third party working for the organisation or performing services for it; a duty owed as an occupier of premises; a duty owed in connection with the supply of goods or services, the carrying out of any construction or maintenance operation or the keeping of plant or vehicles. This definition would seem to include most of the activities covered by Sections 2 to 6 of *HSWA*.

The duty is breached in a manner defined as 'gross' as per Section 1(b) if the conduct alleged 'falls far below what can reasonably be expected of the organisation in the circumstances'. This is a very subjective definition, but it is clarified by Section 8(2), which lays down the criteria a jury should take into consideration when determining if the breach is to be deemed as 'gross'. These factors will include such matters as the seriousness of the failure, how much of a risk of death it posed and the extent to which the evidence produced shows that there were 'attitudes, policies, systems or accepted practices within the organisation that were likely to have encouraged any such failure'.

As such, *CMCHA* is based on Civil Law concepts and focuses on organisational failings in the health and safety management system. There is no reverse burden of proof as in Section 40 of *HSWA*, so the prosecution will have to prove their case by the normal standard for criminal cases, in other words beyond all reasonable doubt. However, there is no individual liability under the act and so the only penalty is a financial one, although it is always possible to bring a prosecution against an individual for common law gross negligence manslaughter.

It is noticeable that the prosecutions taken so far under *CMCHA* have tended to involve smaller organisations rather than large corporations. The first prosecution was of a small company engaged in land surveying and was the result of the death of an employee caused by the collapse of a trench when he was taking soil samples. In *R v Geotechnical Holdings Ltd (2011)*, the defendant company was run by a single director, who at the time of trial was unfit to plead. The company was found liable for Corporate Manslaughter under *CMCHA* and was fined £385,000, the individual manslaughter case having been dropped.

The subsequent case of *R v Lion Steel Equipment Ltd (2012)* was against a medium-sized company following a fatal fall from a roof at of one of their premises. In this case, an action for Corporate Manslaughter was brought against the company, along with prosecution for individual manslaughter and breaches of Section 37 in respect of three company directors. As with the Geotechnical case, the prosecution against the individual directors for both common law manslaughter and the Section 37 offences were both dropped and the company was found guilty of Corporate Manslaughter under *CMCHA*, being fined £480,000.

The interplay of Corporate Manslaughter charges and individual manslaughter charges has proved problematic, since the introduction of Corporate Manslaughter liability under *CMCHA*.[14] In particular, there is evidence that the prosecution may be prepared to drop the individual manslaughter case and to pursue the corporate liability case instead, as such a case may not be as vigorously defended given that there is no custodial sentence involved. As a result, it could be argued that the key result of introducing *CMCHA* has been to replace the indictment of a company for a breach of Section 2 of *HSWA* with an indictment for Corporate Manslaughter.

Conclusion

The criminal liability for health and safety has been subject to a number of key changes over the last few years, some of which have had a significant effect. The *HSOA* is arguably the most important, because it has led to the imposition of custodial sentences for breaches of Sections 2 to 8 of *HSWA*, which were not available before. The greater availability of custodial sentences has led to managers being more aware of the fact that they may end up losing their liberty in the case of an accident at work. Yet so far, the number of people actually imprisoned as a result is very small, so it has not had time to really have an effect. The other factor is the tendency for courts to convict following a serious accident, unless the defendant can prove that there were very compelling reasons why he should not be liable. As a result, liability under *HSWA* has become more result-based rather than fault–based as seen in *R v Chargot (2008)*, which is arguably not how Criminal Law should operate.

The introduction of Corporate Manslaughter liability, while certainly much trumpeted, is arguably less significant than the introduction of custodial sentences for breach of *HSWA*. Although there is the possibility of very high financial penalties being imposed for breach of the act, as yet the fines levied have been quite modest, possibly because no large corporations have been prosecuted. The main effect of the legislation seems to be to have been the change on the indictment, from breach of a *HSWA* to manslaughter.

The greater problems arise from the diminishing role of the HSE in pursuing criminal actions. The recent decision to limit the HSE to proactive inspection of only those workplaces seen as particularly hazardous, means that the threat of an unexpected visit from the inspectorate has much diminished. We are moving from a proactive to a reactive system, which could lead to a reduction in the reporting of less serious injuries. Instead, the role of civil liability is likely to increase with the insurance industry putting real pressure on employers to ensure effective health and safety in the workplace. As a result, it is to the area of civil liability that I now wish to turn.

Notes

1 Lord Robens 'Safety and health at work: report of the committee 1970–1972' (1972) Cmnd 5034 (London: HMSO)
2 David Branson 'The burden of liability' *SHP* June 2007
3 Lofstedt (op cit), Chapter 4, para 17, p. 39
4 Simon Antrobus 'The criminal liability of directors for health and safety breaches and manslaughter' *Criminal Law Review* 2013, 4, 309–322
5 See the case of the electrocution at Fatty Arbuckles restaurant in August 1998 when the safety adviser was prosecuted
6 See J.R. Spencer, 'Criminal liability for accidental death: back to the Middle Ages' *Cambridge Law Journal* 2009, 68(2), 263–265

7 See David Bergman 'Corporate misconduct' (1999) *NLJ* 1849; and Slapper, G. *Blood in the Bank; social and legal aspects of death at work* (Aldershot 1999)

8 Industry Guidance 417: Health and Safety at Work – Leadership Actions for Directors and Board Members (INDG 417) by the Health and Safety Executive and Institute of Directors

9 The case of Simon Jones who was killed at Shoreham docks in May 1998. The HSE and the Crown Prosecution Service initially refused to prosecute and only did so after a judicial review of the case in March 2000

10 Alexandra Dobson 'Shifting sands and multiple counts in prosecutions for Corporate Manslaughter' *Criminal Law Review* 2012, 3, 200–209

11 G. Slapper, 'Corporate manslaughter: an examination of the determinants of prosecutorial policy' (1993) 2 Social and Legal Studies 423, quoted in Sarah Field and Lucy Jones 'Death in the workplace: who pays the price?' *Comp Law* 2011, 32(6), 166–173

12 Report of court No 8074: Public inquiry into the sinking of the Herald of Free Enterprise chaired by Lord Justice Sheen (September 1987)

13 See *R* v *Balfour Beatty Rail Infrastructure Services Ltd (2006) EWCA Crim 1586; (2006) All ER (D)* at 47 at para 24

14 See Antrobus (op cit)

3 Civil liability for Negligence and breach of statute

Introduction

In this chapter, I would like to look at the legal remedies available for persons bringing a civil claim for an injury at work or arising from work activities. The law here has been subject to a number of changes, not least in the last few years. This is because the need for the injured party to be able to claim compensation for his injuries has to be balanced against the danger of imposing excessive liability on employers. Even though most employers must take out insurance to cover employee claims under the requirements of the *Employers Liability (Compulsory Insurance) Act 1969*, it is still the case that employers often have to pay increased premiums following a claim. So while the courts are reluctant to leave an injured employee without means of redress following an injury at work, there is also a feeling that the employer should not be liable without fault on his part.

Until recently there were three main ways in which a claim could be made for compensation by an injured employee: under Contract, under the Tort of Negligence and under breach of statutory duty. The contractual remedy is of least importance and I will discuss it briefly in Chapter 4. The action for breach of statutory duty was very important but this was abolished in October 2013 for all future actions. However, I will look at this area as it is still of importance for cases started before that date. The main method of action now available is the action under the common law Tort of Negligence and it is this that will comprise the main part of this chapter. In addition, I will look at the liability of employers as occupiers under *Occupiers Liability Acts 1957* and *1984*, as well as product liability under the provisions of the *Consumer Protection Act 1987*.

Liability for the Tort of Negligence

Nowadays, the main area of liability for the employer is under the common law Tort of Negligence. This area of law developed in the late nineteenth century as a means of providing a remedy to employees injured at work, who were unable to sue in contract. At first the **tortuous** remedy was limited by the doctrine of 'common employment', which held that an employer was not liable to his employees for the torts of other employees, a principle upheld in *Priestley* v *Fowler (1838)*. However, the courts later began to counter this by developing the concept of a primary duty of care being placed on the employer, as well as limiting the defence of volenti or consent in situations of employment. As a result, in *Smith* v *Baker & Sons (1891)*, an employer was held liable to his employee who was injured by a falling rock in a quarry, and this case began the development of legal liability in this area.

A further problem was the defence of contributory negligence, which originally would deny remedy to a party if he was in any way liable for the accident. However, the *Law Reform (Contributory Negligence) Act 1945* replaced the complete defence with a partial defence, which meant a reduction of damages for the claimant if he was partly liable. From this time on, the employee would only have to prove the existence of a duty of care, the breach of that duty and the fact that his injury arose as a consequence of the breach in order to establish tortuous liability on the part of the defendant. These are the three key elements of the Tort of Negligence, which I will consider below.

The duty of care

The basic elements of the duty of care in health and safety owed to employees were those laid down in the leading case of *Wilsons & Clyde Coal Ltd* v *English (1938)*. This case involved a serious injury in a mine, caused by moving plant, which was in collision with the employee. The case established that the duty of care was personal to the employer and could not be delegated to another employee. The duty of care was owed to a person acting in the **course of employment** and was not strict, but qualified by what was 'reasonably foreseeable'. The duty of care as established here was fourfold: to provide; safe plant and equipment, a safe place of work, competent fellow employees and a safe system of work. These duties have been extended in subsequent cases to include a duty to provide adequate supervision and the duty to provide a safe working environment. It may be noted that these duties provided the basis for the statutory duties in Criminal Law to be found in Section 2(2) of the *Health and Safety at Work etc. Act 1974 (HSWA)*.

Duty to provide safe plant and equipment

The duty here is to provide safe and adequate plant and equipment for use at work. The duty will be breached if the employer fails to supply any equipment at all, as in *Williams* v *Birmingham Battery & Metal Co (1899)*, or where the equipment supplied is not properly maintained. The liability in Negligence is expanded upon in the *Provision and Use of Work Equipment Regulations 1998 (PUWER)*, and in the past an action would be brought concurrently in Negligence and for breach of the statutory duty under

PUWER. As a result, much of the relevant case law here refers to the statutory duty but would still be valid in defining the liability under Negligence. So, for example, the courts have had to determine the definition of the term 'work equipment'. This has been widely defined to include not just the obvious tools of the trade, but also a van in *Bradford v Robinson Rentals (1967)*, as well as a kerbstone used to push other kerbstones into position as in *Knowles v Liverpool City Council (1993)*. In fact the term 'work equipment' need not be something actually used to do the work but can include anything performing a useful function in the workplace such as a self-closing door mechanism, as in the more recent case of *Spencer-Franks v Kellogg Brown & Root Ltd (2008)*.

The employer's liability is not just for the equipment he provides and maintains, but also in respect of any defective equipment provided by a third party, which he makes available to employees for use at work. The legal position was clarified by the *Employers Liability (Defective Equipment) Act 1969*, which made the employer primarily liable for any equipment provided for use at work. This overruled the previous common law position, whereby the injured party was required to sue the supplier of the equipment, as in *Davie v New Merton Board Mills Ltd (1959)*. The primary liability of the employer is based on the fact that the employer is required to take out insurance against liability for injury to his employees, under the *Employers Liability (Compulsory Insurance) Act 1969*.

Duty to provide a safe place of work

The duty to provide a safe place of work in common law is similar to the statutory duty of the occupier to lawful visitors under the *Occupiers Liability Act 1957* (see p. 73). As such, this is the civil law equivalent of the criminal liability under Section 4 of *HSWA 1974*. Until October 2013, it was possible to bring a joint action under Negligence and the *Workplace (Health, Safety & Welfare) Regulations 1992 (WHSWR)*, and some of the terms defined by the court were in respect of the statutory duty, even though applicable in the common law action.

In defining the term 'place of work', the courts have taken a wide interpretation, as for work equipment. As a result a place of work can include a ship as in *Coltman v Bibby Tankers Ltd (1988)*, the client's premises as in *General Cleaning Contractors v Christmas (1953)*, or even as vehicle used to ferry a person back from a public house during an army training weekend as in *Jebson v Ministry of Defence (2000)*. Sometimes it is difficult to distinguish between a place of work and work equipment, as in the case of vessels or vehicles, which may fall into either category.

The employer is required to ensure that a place of work is safe from any reasonably foreseeable risks. As such, the liability is not strict, so the employer was not liable in *Latimer v AEC (1953)* when the employee slipped on wet sawdust while trying to clear away rainwater, as the court felt that the employer had taken all the precautions that he could. However, for cases brought under the statutory duties, the courts have been prepared to take a stricter view. This can be seen in the case of *Ellis v Bristol City Council (2007)*, where the council was deemed to be liable for a breach of Section 12(1) of *WHSWR*, when the claimant slipped on a pool of urine in a nursing home run by the defendants. It was held that the employer had a duty to ensure that the floor surface was 'suitable' for the purpose for which it was to be used, and this duty was not qualified by what was 'reasonably practicable' but was effectively strict. With the demise of the statutory remedy, it remains to be seen whether the court will continue with this stricter interpretation or revert to their earlier approach as seen in *Latimer*.

Duty to provide a safe system of work

The term 'system of work' is very wide and includes the planning of a work operation, as well as the method by which it is carried out. This means that it will include the way the workplace is laid out, the training and supervision involved, the provision of personal protective equipment and the setting up of emergency arrangements. So for example, the creation of a task that involves repetitive actions and exposes the employee to repetitive strain injury, will be seen as evidence of a poor system of work. This was the case in *Mountenay (Hazzard) & Others v Bernard Matthews (1993)*, where the problem arose from the processing of poultry, which involved repeated arm and wrist movements leading to repetitive strain injury. Similarly, the requirement of bank staff to input large amounts of data onto a keyboard, involving repeated arm and wrist movements, was also seen as an unsafe system of work, as in the case of *Alexander & Others v Midland Bank PLC (2000)*.

In addition to the way in which the work is performed, an employer would be liable if he failed to warn employees of the dangers involved in a particular activity, such as any health effects created by the use of hazardous substances. So in *Pape v Cumbria County Council (1991)*, the employer was in breach of his duty of care because he failed to inform the employee of the dermatitis hazard involved in the use of the cleaning material provided. However, we should note that the liability here is not absolute, so the employer would not be liable if it was possible for the employee to avoid the risk by asking for extra help. In *Chalk v Devizes Reclamation Co. Ltd. (1999)* the employee was injured when he undertook a manual handling operation involving an unusually large piece of metal scrap. The employer successfully argued that the employee had been fully trained in manual handling techniques for normal loads. However, as this was an unusual load, he believed that the employee should have asked the employer for advice before trying to lift it.

We can see that this duty is very vague and can be expanded to cover a wide range of activities. The key requirement is that the employer needs to give thought to the nature of the operation and the safety arrangements provided. It is very important to carry out a suitable risk assessment, and in many cases this would lead to the creation of a formalised safe system of work, which should be communicated to the employee through effective training and supervision. Where the dangers are quite acute or the hazards are multiple, it may be necessary to provide a permit to work system, such as for access to confined spaces or working on live electrical equipment.

Duty to ensure competent fellow employees

The duty here is on the employer to ensure that his employees are able to carry out their tasks competently and safely, and to take action against any employee who is seen as posing a risk to fellow employees or third parties. This duty incorporates both training and supervision, and is therefore the Civil Law equivalent of the criminal duty of care in Section 2(2)(c) of *HSWA 1974*. The need to provide effective training is central here, especially where the employee is exposed to new hazards as a result of changing processes or the use of new materials.

In addition, an employer has been held liable if he fails to control a well known practical joker who then causes injury to another employee, as in *Hudson* v *Ridge Manufacturing Co. Ltd (1957)*. On the other hand, in *Smith* v *Crossley Brothers Ltd (1951)*, the employer was not held liable for a single unexpected action by a practical joker that injured an employee. These earlier cases are quite old and I would suggest that the duties of the employer here have been extended so that he would now be liable for any injury caused if he has not exercised adequate supervision.

The issue of effective supervision is related to the question of which actions are deemed to be carried out in the course of employment, as an employer would not normally be liable for action outside of this. However, if an employee disobeys his employer's instructions, this does not mean he is acting outside of the course of employment, as this would apply to many, if not most accidents. So the employer was liable for his employee's actions in *Kay* v *ITW (1967)*, where a fork lift truck driver injured a third party while reversing a goods vehicle in breach of company instructions. It would seem that the only situation where the employer would not be liable is where the activity is completely unrelated to his job, and even then, if the injured party is another employee, he too must have been complicit and also acting outside of the course of employment. The age of the employee is arguably of importance here, as the younger and less experienced he is the greater responsibility will fall on the employer to provide adequate supervision.

Other duties on the employer

Following on from the case of *Wilson & Clyde Coal* v *English (1938)*, the courts have developed the extent of the duty of care in various ways. This will now include the need to provide a safe working environment as well as adequate welfare facilities, such as the provision of places for washing and changing as in *Mc Ghee* v *NCB (1973)*. The duties are usually qualified by the requirement of what is 'reasonable foreseeable', so that the employer is not liable for hazards that it would be unreasonable for him to be aware of at the time. This can be seen in the case of *Baker* v *Quantum Clothing Ltd (2011)*, where the employer was held not to be liable for industrial deafness caused by high noise levels in his factory, where there was uncertainty at the time as to how low the safe noise level should be.

The nature of the injury

It has long been the case that the employer is liable for physical injury resulting from traumatic events, and in later times the liability has also extended to **latent injury** such as asbestosis and **silicosis**. However, in both cases the injury is physical and the main problem is one of causation, especially in the case of latent injury. In recent years, the key development has been the extension of liability to include purely mental injury. This in itself can be subdivided into two main types of injury, namely mental breakdown and post-traumatic stress disorder, the latter usually referred to by lawyers as 'nervous shock'.

Mental breakdown

In the case of mental breakdown injury, the leading case is *Walker* v *Northumberland County Council (1995)*. This case involved the failure of an employer to deal with the excessive workload to which the employee was subjected. The claimant suffered a nervous breakdown, which led to him being off work. When he returned he discovered that the workload was even greater and within six months had suffered a more serious breakdown, which caused him to stop work permanently. The employers were liable for his injury because they had been made aware of the problem and yet failed to take action to deal with it.

This case poses problems, because mental stress may be difficult to detect, especially if the employee is trying to hide it. Also the employer may feel that the employee has brought the problem on himself by taking on too high a level of responsibility where higher stress levels may be expected. Subsequent cases have clarified the legal position, making it clear that the employee has to bring the stress problem to the attention of his employer, otherwise the injury may not be seen as 'reasonably foreseeable'. In this respect the employer is entitled to assume that the employee is able to withstand the normal pressures of the job, unless the contrary is either expressly brought to his attention or is abundantly clear from the employee's behaviour or sickness pattern, a point made clear in the case of *Sutherland* v *Hatton (2002)*.

The case of *Barber* v *Somerset County Council (2004)* established that the employer had a duty to investigate the cause of an employee's stress following an absence from work, and that failure to do so meant that he would fall below the standard of care that could be expected of a reasonable and prudent employer. Subsequent cases such as *Hartman* v *South East Essex Mental Health & Community Care NHS Trust (2005)* also made it clear that the employer has a greater need to monitor the condition of the employee where the employment is in a high-risk occupation.

Where it is clear that the employer is aware of the stress-related illness of the claimant, he is liable for any reasonably foreseeable consequences. This could even include the death of the employee if this is the end result. So in *Corr* v *IBC Vehicles (2006)* the employer was liable for the suicide of the employee, which was the result of depression caused by a physical injury at work. This case illustrates the fact that the courts do not draw a real distinction between physical and mental injury.

Nervous shock

As regards so called 'nervous shock', this is basically the mental reaction to a traumatic incident, which does not involve any physical injury to the claimant, although it will often involve physical injury or death to a third party. In law, a distinction is made between 'primary' and 'secondary' victims. A primary victim is one who is directly threatened with death or injury, so that the mental trauma is a result of extreme fear for his own safety. In this case the claimant need only prove the normal requirements of Negligence, in other words that the injury was reasonably foreseeable. The mental injury here is seen as a direct consequence of the fear of physical injury and so is claimable, as in the case of *Dulieu* v *White (1901)* where the claimant believed she was about to be killed by the defendant's runaway horse drawn van when it crashed through the window of the public house where she was working.

Where the claimant was never in actual physical danger, he will be seen as a secondary victim. Such a person can only bring a claim if he can prove that he is in close physical proximity to the accident and that he can establish a relationship of 'close love and affection' with the injured party. This second requirement will normally exclude most employment cases, as a claimant will not usually be able to argue such a close relationship with a work colleague. This was made clear in the case of *Robertson & Rough* v *Forth Road Bridge Joint Board (1995)*, where the employer was held not liable for the mental trauma caused to the claimants, who witnessed the death of their colleague at work when he fell from the Forth Road Bridge.

It may be possible to claim damages if the employee himself is led to believe that he is the cause of the accident and the subsequent death or injury when this is actually the fault of the employer, a situation which occurred in the case of *Dooley* v *Cammel Laird (1951)*. However, the employer would not be liable if the employee was never actually in fear of his life and did not have any reason to believe that he was responsible for the death or injury of others. An employee is expected to demonstrate a certain level of fortitude in the case of danger and this may defeat such a claim. This was the case in *McFarlane* v *EE Caledonian (1994)*, where the claimant suffered nervous shock as a result of working on a rescue ship in the Piper Alpha accident. He argued that he was traumatised by a large explosion that looked as if it would engulf the ship, but it was held that his life was never seriously in danger and that no liability could ensue.

The breach of the duty of care

Once a duty of care is established, it is then necessary to show that the employer has breached that duty. In this respect the standard of care used is that of the 'reasonable man' or as regards the workplace, the 'reasonable employer'.

The standard of care

In determining what constitutes the actions of the 'reasonable employer', several factors will be taken into consideration.[1] First, the knowledge of the employer, or at least the knowledge that the employer could 'reasonably' be expected to have in the circumstances. In *Stokes* v *GKN Ltd (1968)* it was held that the employer had to follow current work practices, but also be aware of any new developments in health and safety. However, an employer does not have to have total foresight and would not be expected to be aware of every new recommendation, especially long before it became law.

This was the basis of the decision of the Supreme Court in *Baker* v *Quantum Clothing PLC (2011)*, where the employer was held not liable for the hearing loss caused to employees, working in a factory where the noise was over 80 decibels, when the legal limit at the time was only 85 decibels, although there had been discussion as to the need for a lower limit. Compliance with the existing regulatory requirements was seen as satisfying the common law standard of care.[2]

In addition to the knowledge of the employer, the level of risk involved is important. In this respect, civil liability mirrors the importance of risk in criminal liability in establishing what is 'reasonably practicable'. Therefore, where an employee is seen as being more at risk, because he is disabled, young or untrained, then a greater duty of care lies on the employer, which means the duty is more easily breached. The leading case here is *Paris* v *Stepney Borough Council (1951)*, where the employer was liable for the injury to a partially-sighted employee who had failed to wear eye protection. Even though such eye protection was not usually worn to do the particular job involved, the employer was under a greater duty to ensure that eye protection was worn in this case, as the employee had only one good eye.

In addition, the greater the risk the work operation involves, the more stringent the precautions required. Therefore, if the risk is seen as very low, the precautions taken may be minimal and the employer would have no liability for a freak accident. This can be seen in the case of *Latimer* v *AEC (1953)* as discussed above. The employer argued that he had done everything a reasonable employer could do to prevent an accident, and indeed the employee was taking action to avoid an accident happening to the rest of the staff. Moreover, in the more recent case of *Stark* v *Post Office (2000)* the court did not find liability in Negligence for what was arguably a freak accident and the case had to be pursued under breach of statutory duty. Consequently, it is still the case that there is no liability in Negligence where there is no fault on the part of the employer.

The burden of proof

In all civil actions, the burden of proof lies on the claimant to prove his case. The standard of proof required is 'the balance of probabilities', which effectively means more than a 50 per cent likelihood that the claimant is justified in bringing his claim. This is a much lower standard than the criminal standard of 'beyond all reasonable doubt'. Nevertheless, it can still be difficult to prove liability where the cause of the accident is uncertain or it is the result of a multiplicity of causes. To help the claimant, there are two situations where the burden of proof may be switched to the defendant.

First, the legal concept of '**res ipsa loquitur**' may apply, meaning that 'the facts speak for themselves'. This can be used where there is no logical explanation for the accident, if the accident could only have come about by negligence on someone's part, and where the defendant had full control over the situation that led to the accident. In this case the burden of proof is switched to the defendant, and he must show how the accident could have happened without negligence on his part in order to escape liability. One of the earliest cases involved an accident at work in *Scott* v *London & St Katherine's Docks (1865)*, where the claimant was injured by a bag of flour, which fell from an opening on one of the upper floors of the defendant's warehouse. The defendant was held to be liable as he could not explain how this could have happened unless there had been negligence on his part.

The second situation arises once a defendant is convicted of an offence arising from the same circumstances, as have given rise to the civil action. Under Section 11(c) of the *Civil Evidence Act 1968*, it is possible for the fact of the conviction to be used in evidence in the subsequent civil action. This is a key reason why the claimant will wait for any criminal action to complete its course, before pursuing this civil case. Moreover, the fact that a criminal conviction is based on a higher standard of care than that in civil action means that such evidence will carry greater weight in the subsequent civil case.

Damage arising

In addition to proving a duty of care and breach of the duty, it is also necessary to show that the damage arose as a result of the breach. This brings us to the legal concept of causation, which itself can be divided into two types: factual causation and legal causation or **remoteness**.

Factual causation

This is the requirement to show that the accident was the actual result of the breach of the duty of care. In the Tort of Negligence, the main test is the 'but for' test, in other words the accident would not have happened 'but for' the acts or omissions of the defendant. This means that the claimant cannot win his case if the evidence is that the accident or injury would have occurred regardless of the breach of duty by the employer. There are a number of problem areas here as set out below.

CONSENT OF THE CLAIMANT

In some respects this overlaps with the defence of 'volenti' or consent, in as much as the cause of the accident is attributed solely to the actions of the claimant. So in *McWilliams* v *Sir William Arrol & Co. Ltd (1962)* the claimant was the widow of the deceased employee who had fallen to his death from a crane. The employer had originally provided safety harnesses on site but had later withdrawn them shortly before the accident. However, the evidence was that the deceased had never worn the harness and was unlikely to have done so. Nowadays we might argue that the employer should have exercised more effective supervision and not allowed the employee to ascend the crane without the safety belt. However, if we accept that such supervision is not always effective, it might be possible to argue that the employee was wholly responsible for the accident, or that there was no link between the breach of duty of the employer and the accident to the employee.

EXISTING MEDICAL CONDITION

A rather different problem arises in the case of illness or work-related disease, as these may be multi-causal and so it may be difficult to determine whether the illness was the result of the employer's negligence or a pre-existing medical condition. This can lead to some unfortunate decisions for the claimant, as can be seen in *Jobling* v *Associated Dairies (1982)*. Here the claimant suffered an injury at work due to the negligence of his employer, which reduced his earning capacity by 50 per cent. However, before the case came to court, he developed a wholly unrelated back problem, which eventually rendered him totally unable to work. The employer was held liable for the reduction in earning capacity, but only up to the time when the back problem stopped the claimant being able to work. It is to be noted that had the case been settled before the later back condition came to light, the compensation would have been paid to cover the whole of the remainder of his expected working life.

However, if the second event had been a non-natural event, such as a negligent or criminal action, then the employer would have been liable for paying the compensation for the rest of the claimant's working life in respect of the injury he had caused. This was made clear in the case of *Baker* v *Willoughby (1970)*, where the claimant suffered an ankle injury at work due to the negligence of the defendant, but subsequently lost his leg as the result of a criminal assault by a third party. In these cases, the courts are concerned not to award compensation for the mere vicissitudes of life, but they will not limit compensation where the subsequent action is non-natural.

LATENT INJURY

A more serious problem arises from those situations where the illness can be linked to an employment activity, but it is not possible to attach specific liability to a particular employer. A case in point here is the disease of mesothelioma, which is caused by the inhalation of asbestos fibres. Unlike asbestosis, which is the result of cumulative exposure, mesothelioma may result from the inhalation of a single fibre. If an employee has worked for a number of employers, all of whom have failed to prevent exposure to asbestos, it is impossible to determine which employer is responsible for the fatal exposure. This problem was raised in the leading case of *Fairchild* v *Glenhaven Funeral Services Ltd (2002)*, in which the House of Lords decided on policy grounds that all of the employers for whom the claimant had worked would be **jointly and severally liable**. As a result, damages would be paid out in proportion to the amount of time the claimant had worked for each party, but all parties were still liable for the whole amount if necessary. This decision clearly modifies the legal principle of causation, as it is impossible to determine liability on the part of any one employer on the civil basis of 'the balance of probabilities'.

However, in the subsequent case of *Barker* v *Corus (2006)* it was held that imposing joint and several liability was unfair, because some of the parties were not traceable at the time of the action, so the remaining parties would be liable for the whole of the damages. Instead, the court said that the liability should only be several, so that parties would only be liable for a proportion of damages related to the amount of time the claimant had been employed by them. This meant that claimants might not obtain full compensation, where some of the employers were untraceable by virtue of going into liquidation, putting the loss on the employee's shoulders.

This problem was swiftly dealt with by the *Compensation Act 2006,* which stated that the liability would be joint and several,[3] meaning that the claimant can sue any or all of the parties for the whole amount. It is then up to the defendants to claim a **contribution** from what other parties they can trace. In effect, the burden has been shifted from the employee to the employer, or rather his insurance company.

We can see that the area of causation has created some problems in respect of injury at work, especially latent injury, where the cause is difficult to clearly identify. The law here has been driven by the desire to ensure that the claimant will always have a party to sue for any injury caused at work, given that the employer will have an insurance policy to cover such an event. The aim has been to place the burden on the widest shoulders, although this may cause some problems for employers. Even then, the impact of subsequent diseases can lead to unfair results. The problem of industrial disease is a growing one and is responsible for an increasing share of compensation payments for insurance companies. For this reason, it is clear that this will pose a continuing challenge to the entitlement of injured parties over the next few years. We may have to question whether a purely fault-based system can really deal with these issues, in an area of law where it may be difficult or impossible to allocate fault in the first place.

Legal causation

Even when it is possible to prove factual causation, the law will sometimes limit liability for policy reasons, in order to prevent liability becoming too extensive. This is the so-called 'floodgates' argument, whereby the law seeks to avoid defendants being subject to an indeterminate liability from an indeterminate number of claimants for an indeterminate time. In this respect, the law requires that the loss be 'reasonably foreseeable', in much the same way as these terms limit the duty of care itself. As such, the Tort of Negligence will not allow an employee to claim for a completely unforeseeable event at work, such as a freak accident.

THE EGG-SHELL SKULL RULE

However, in respect of personal injury, the law is modified by a policy based decision that favours the claimant, called the 'egg-shell skull rule'. This rule, which can also be found in criminal law, states that once it can be shown that some injury was 'reasonably foreseeable' as a result of the defendant's negligence, the defendant is liable for all of the consequences of the accident, however unforeseeable. So in *Smith* v *Leech Brain & Co Ltd (1962)* the claimant's husband was splashed on the lip with molten metal, but this triggered a cancerous growth, which eventually killed the employee. As the burn injury from the splash was reasonably foreseeable, the defendant was liable for the subsequent fatality. In effect, the defendant was taking responsibility for the fact that the claimant might have a specific medical weakness that increased his susceptibility to further injury.

This can apply even where the cause of the further injury is due to the action of a third party, so long as that action was not negligent in itself. In *Robinson v Post Office (1974)* the employee slipped on a ladder and cut his leg, requiring him to go to hospital for treatment. At hospital he was given a tetanus injection, to which he reacted badly and suffered brain damage. The Post Office was held liable for the brain damage, even though they could not have known that he would react in this way and had acted as any proper employer should have done. Once again, we can see that the law seeks to put the liability on the employer in order to enable a claim to be made against the insurance policy. The alternative would be to allow the employee to bear the burden, or more likely to put the burden onto the state.

Liability to employees

Liability to labour-only sub-contractors

In addition to the employer owing a duty of care to his own employees, he may also have a similar duty to contractors who are engaged by him, but are effectively under his control. This is to prevent employers using bogus 'self-employment' as a means of avoiding liability. The leading case here is *Ferguson v John Dawson (1976)*, where a self-employed construction site worker was injured while at work. The court held that he was effectively an employee working for the party who engaged him, in respect of any liability for health and safety. A later case that confirmed this approach was *Lane v Shire Roofing Ltd (1995)*, where again the claimant, who was nominally self-employed, was seen as an employee for health and safety purposes. The court is aware that the employer is the party in the best position to take out relevant insurance and that it is unfair to put the liability on the worker, who is less likely to be able to insure himself.

Liability for agency workers

The position is rather different where the employee is loaned to the employer by an agency. The primary employment relationship here remains with the agency as made clear by the *Agency Workers Regulations 2010*. As such, the agency has the main responsibility for the health and safety of the agency worker and will be civilly liable in the case of an accident. However, the agency will be able to claim compensation from the **end user** if the latter is responsible for the accident. This clarifies the previous position, where agency workers could be seen as neither employees of the agency or the end users, but effectively self-employed. This was the position arrived at in the case of *James* v *Greenwich Council (2007)*, a case which demonstrated the unsatisfactory legal situation prior to the passing of the *Agency Workers Regulations 2010*.[4]

Liability for employees

Liability for employees to third parties

The employer will not only be liable to his employees for their injuries, he will also be liable to third parties for injuries cause by his employees. In this respect, his liability is no different from that to his own employees. The employer is vicariously liable for the torts of his employees, so he will be liable for common law Negligence under the principles laid down in *Wilson & Clyde Coal* v *English (1938)*.

The employer will be liable for the action of any employee, which is carried out 'in the course of employment'. This term is widely defined by the court and will include any action carried out for the benefit of the employer. As such, it will include actions that are both unauthorised and illegal, such as the moving of a lorry by a fork lift truck driver in *Kay* v *ITW (1967)*, as discussed above.

Liability for employees loaned to third parties

This situation is similar where an employee is loaned out by one employer to another. Where the new employer exercises effective control over his actions, then the new employer bears the liability for his health and safety as in *Garrard* v *Southey (1952)*. Only where the loaned employee is not under the control of the new employer is the situation different, in which case responsibility lies with first employer. This was the case in *Mersey Docks and Harbour Board* v *Coggins and Griffiths (1947)*, where the employer hired out a crane driver to a stevedore company, when he negligently injured a third party. The House of Lords held that the hirer always retained control over the crane driver and so he was liable for the injury. A similar decision was arrived at in the later case of *Morris* v *Breaverglen Ltd (1993)* involving the loaning out of a dumper-truck driver. Clearly there will always be a dispute as to the level of control that must be exercised by the second employer for liability to pass on to him.

Shared liability

At one time it was thought that only one party could be vicariously liable for the tort of another, usually the direct employer. However, the case of *Viasystem (Tyneside) Ltd* v *Thermal Transfers (2005)* made it clear that liability can be apportioned to two or more parties. In this case, the person who caused the damage was an employee of a subcontractor, but the court held that liability should be shared between his employer and the party engaging the subcontractor, as both parties had an element of control over his actions. This decision means that in complex subcontracting arrangements the legal liability could be shared by a number of parties where the control is also shared. The legal implications of this case are not yet worked out, but it may be that contracting parties could use their contractual agreement to clearly define where the liability lies, so as to ensure that the relevant party can obtain adequate insurance cover. Note here that the courts may seek to prevent this being used, to transfer liability onto a party unable to obtain such insurance, such as a small labour only subcontractor.

Defences to Negligence

Denial of liability

The first defence to Negligence is a simple denial of liability itself. The Tort of Negligence requires the claimant to prove three main elements: the existence of a duty of care, breach of the duty and damage arising. Therefore a denial of liability means denial of one of the elements of the tort.

As regards the duty of care, this is limited to matters that are 'reasonably foreseeable' to the defendant. In respect of the employer's liability, it is also necessary to prove that the claimant is an employee, and that he is acting in the course of employment when he was injured. We have seen that the definition of employment may be widened to include so-called self-employment. In addition, the definition of the course of employment is wide and will include foolish and even unauthorised actions, as in *Kay v ITW (1967)* (see p. 57). However, where the action is completely unrelated to the activities of the employer, the action may be seen as outside of the course of employment so that no duty will lie. This will include such matters as fighting at work, as illustrated in the case of *Coddington v International Harvester (1969)*, where the claimant was set alight when two young employees began throwing flammable thinners at each other.

In addition, no liability will arise where the employer can argue that he was not in breach of his duty of care. The standard of care expected is that of the 'reasonable employer', so this defence will revolve around what the court feels is the relevant standard. If the court sets too high a standard, then this may mean that the liability is more akin to strict liability. In addition, the defendant may argue that there is no causal relationship between the breach of duty and the damage to the claimant, although we have seen that the rules on causation can be amended to prove liability on policy grounds, as in the case of the mesothelioma victims.

Volenti or consent

In addition to negating the key elements of the Tort of Negligence, there are a number of other specific defences, the first of which is 'volenti' or consent.

This means that the claimant is responsible for his own injury, as a result of his acts or omissions, so that the employer is no longer liable. In employment relationships, the law limits the defence of 'volenti' in various ways.

First the consent must be real, so an employee does not consent to the inherent dangers of his job, even if the job is paid a premium rate in recognition of the dangers involved. This principle was established in the case of *Smith* v *Baker (1891)*, where the employer was injured by falling rocks when working in a quarry.

Second, the fact that the employee is aware of the risks still does not mean that the defence of 'volenti' applies, as the court will expect the employer to supervise the employee and ensure that he does not engage in dangerous activities. 'Volenti' will probably only apply in the case of wilful disobedience by an experienced employee, such as can be seen in the case of *ICI* v *Shatwell (1965)*, where two experienced shot firers detonated a charge without taking proper cover, in breach of both work rules and statutory regulations. However, there are very few recent cases to support this approach.

Contributory negligence

Although the courts are usually unwilling to find 'volenti' on the part of the employee, it is always possible for the employer to use the defence of contributory negligence, where the claimant is partly responsible for the accident. Under the provisions of the *Law Reform (Contributory Negligence) Act 1945*, if the employee is seen as partially liable, damages will be reduced as a consequence. However, for contributory negligence to apply there must be some fault on the part of the employer, otherwise it would be possible to argue that there is no breach of the duty of care by the employer.

Act of a third party

Usually this would be a complete defence to Negligence. However, in the case of an employment relationship, the employer is primarily liable for any injury to his employee while in the course of employment, although he can claim a contribution from a third party if necessary.

Where injury is caused by a third party, the employer will claim against that third party such as another employer or even a member of the public. However, it also means that he can claim a contribution from an employee, if he is responsible for the injury to another employee. This principle was established in the case of *Lister* v *Romford Ice & Cold Storage Ltd (1957)*, an unusual case involving a father and son who both worked for the same employer. Here the son was responsible for reversing a vehicle over his father, and although the father was able to recover compensation the court held that the employer could recover the compensation from the son. In general, employers will be reluctant to take action against an employee, as this is likely to undermine the relationship with other employees. However, there is no reason why a contribution may not be sought, and this may be more likely where the employee has engaged in an action that is deliberate or highly reckless.

Limitation of actions

An action for personal injury under the Tort of Negligence must be commenced within three years of the injury occurring, or the action will be struck out by the courts as being out-of-time or time-barred. This is to prevent a miscarriage of justice due to reliance on evidence in respect of an incident, which may be many years in the past and where witnesses may no longer be able to remember the facts correctly.

While the limitation period is not usually a problem for actions in respect of **traumatic injuries**, it does pose a major problem for latent injuries such as diseases like asbestosis or mesothelioma. As a result, Section 14 of the *Limitation Act 1980* allows for an action to be brought up to three years after the claimant is aware of the disease and is aware of the identity of the party who is responsible. This means that actions may be brought many decades after the person worked at the defendant's workplace, hence the need to keep extensive records in respect of environmental monitoring and health surveillance.

Moreover, under Section 33 of the *Limitation Act 1980* it is always possible for the court to waive the application of the limitation period if they think there is a good reason, such as evidence of deliberate deception on the part of the defendant. We should also note that in the case of a fatality, there is a fresh limitation period that runs from the date of the death or knowledge of the death, both of which may occur some time after the time of the original injury.

Act of God or inevitable accident

This is not really a very effective defence as regards the liability of employers for their employees. The general view is that the duty of care of the employer extends to any reasonably foreseeable event, and so the claim that an accident is an act of God is effectively arguing that it is not foreseeable and so the duty of care does not apply to such matters.

There are few examples of employees being denied a remedy for an accident at work. A possible example would be *Lewis* v *Avidan (2005)*, where the claimant slipped in a pool of water at the employer's premises, but it was held that it was not 'reasonably practicable' for the employer to have prevented the accident that was caused by a sudden water leak. While it is the case that the courts will not provide compensation for a freak accident as such, they are likely to argue that there was a lack of adequate supervision by the employer, so as to avoid sending the employee away without any compensation

Breach of statutory duty

Development of the duty

The ability of an employee to sue for breach of statutory duty for an accident at work dates back to the case of *Groves* v *Lord Wimborne (1898)*. This was an action brought under the *Factory and Workshop Act 1878*, which required dangerous moving parts of machinery to be securely fenced. The claimant was severely injured by an unguarded machine and was able to claim compensation, the court dismissing the defence of common employment on the grounds that the statute imposed an absolute duty on the employer to protect the employee.

Most health and safety legislation passed in the twentieth century allowed a claim for breach of statutory duty, most notably the regulations passed to implement the European Framework Directive on Health and Safety, such as *PUWER* and *WHSWR*. In most cases, a 'double-barrelled' action would be taken, for common law Negligence and breach of statutory duty under the appropriate regulations. However, this has now come to an end following the passing of the *Enterprise and Regulatory Reform Act 2013*, which abolished the right of action for breach of statutory duty in any case brought after 1 October 2013. However, I wish to outline the key elements of the action, because such cases will continue to be brought for some time, until the limitation rules begin to apply.

The advantages of the action

The main advantage of the action for breach of a statutory duty was that the liability was not always fault-based, but could involve strict or **absolute liability**. Therefore, while many statutes are qualified by the term 'reasonably practicable', some impose strict liability. This will include Regulation 5 of *PUWER*, which requires equipment to be maintained 'in an efficient state, in efficient working order and in good repair'. As a result, the employer may be liable to the employee for injury caused by a freak accident of such a type for which they would not be liable in common law Negligence. This was the case in *Stark* v *Post Office (2000)*, where the employer was held to be liable for injuries caused by a sudden failure of a bicycle calliper, due to a defect that could not be detected by reasonable means.

Requirements of the action

In order to take an action for breach of statutory duty, it is necessary to establish three things: that there was a duty actionable under statute, that there was a breach of that duty and that the damage arose as a consequence. The last two requirements are the same as for common law Negligence, but the first requirement is specific to the action for breach of statutory duty.

For a duty to exist under statute, it has to be shown that the statute was designed to provide a right of action for that situation. Moreover, it is necessary to show that the statute was designed to protect that category of persons who are bringing the action. As regards health and safety legislation, the statutes usually cover both employees and self-employed workers, as both are covered by the overall term of 'worker'. However, they will not cover members of the public, as was made clear in the case of *Donaldson* v *Hays Distribution Services Ltd (2005)*, where a customer was unable to claim for injury caused by a reversing lorry under the provisions of *WHSWR*, as these regulations were passed to protect workers.

Criminal statutes relating to health and safety have long been seen as imposing civil liability on employers, except in specific cases such as *HSWA* or parts of *MHSWR*. This is because the purpose of the health and safety regulations is to protect persons at work, and a right of civil action was seen as a way of helping to do this by holding the employer to account for any health and safety failings. The extent of the duty is quite wide and will include any injury to employees or workers, even if the injury does not occur on work premises, so long as the person was acting in the course of employment. This can be seen in the case of *Fraser* v *Winchester Health Authority (1999)*, where the claimant was injured by a camping stove, which he had not been properly trained to use.

Problems with statutory duty

The main problem is that the statute must be applicable to the circumstances, so that the interpretation of the terms may be subject to dispute in the courts. A good example here is *PUWER*, which imposes strict liabilities on the employer in respect of the provision and maintenance of work equipment in 'an efficient condition'. Much of the case law revolves around the definition of what is work equipment, as opposed to the workplace or a work piece. So, in *Hammond* v *Commissioner of Police for the Metropolis (2004)*, the court held that a nut on the wheel of a vehicle under repair was a work piece and not work equipment, so that it was not covered by *PUWER* and the claimant could not use that legislation to take action in respect of injury caused when the nut sheared. However, a more recent decision in *Spencer Franks* v *Kellog Brown & Root (2008)* held that a door-closing mechanism, which was being worked upon, was work equipment and *PUWER* would apply, effectively widening the interpretation of the term 'work equipment'.

In addition, it is necessary to show that the breach of the statute led to the injury, and this has also caused some interesting problems. So in *Close* v *The Steel Company of Wales (1961)* the claimant was injured by a piece of machinery that was ejected from a drill, and sued for breach of what was then Section 14 of the *Factories Act 1961*. The court determined that the purpose of the guarding under the legislation was to prevent a person coming into contact with dangerous moving parts, not those parts coming into contact with him, so there was no liability on the employer. This gap in the law was later remedied by Regulation 12 of *PUWER 1998*.

In addition, as with common law Negligence, it is necessary to show that the breach of the statute was the cause of the accident or injury. This was the position in *McWilliams* v *Sir William Arroll (1961)* as seen above, where the death of the crane operator was held to be due to his unwillingness to wear a safety harness, rather than the failure of the employer to provide it on the day of the accident.

The other defences to a breach of statutory duty are as for common law Negligence, except that 'volenti' is usually disallowed in an action for breach of statutory duty, unless the employee's actions cause the employer to be in breach of the statute. This was the case in *Ginty* v *Belmont Building Supplies Ltd (1959)*, where the employer provided crawling boards for use on a defective roof, but the employee failed to use them with the result that he fell through the roof. The failure of the employer to comply with the relevant statute was due entirely to the actions of the claimant, so the employer was not liable.

Occupiers' liability

Introduction

An employer is subject to occupiers' liability if he is an occupier of premises, both under common law Negligence and under the *Occupiers' Liability Acts of 1957* and *1984*. However, the liability under statute relates to injury caused by the premises and not by activities that take place on the premises, although it can be difficult to distinguish between the two. Therefore in *Ogwo* v *Taylor (1987)* a burn injury that occurred to a fireman fighting a fire on the defendant's premises was held to be actionable in Negligence, because it was caused by scalding steam created by a fire caused by the defendant's negligence. Common law Negligence here is a fallback liability, which applies whenever the statutory liability is inapplicable. Alternatively, an action in Negligence may be used where there is a long history of using such actions, as in the case of accidents caused by the negligent use of vehicles, which in theory could be brought under occupiers' liability.

The two relevant statutes cover different types of claimant. The *Occupiers' Liability Act 1957 (OLA 1957)* is limited to lawful visitors, although this includes persons allowed on the premises by law such as Factory Inspectors. The *Occupiers' Liability Act 1984 (OLA 1984)* deals with trespassers, but will also cover persons exercising a statutory right of way or a private right of way. It is not always easy to distinguish between trespassers and lawful visitors, as a trespasser may become a lawful visitor if they are led to believe that they have a right to come onto the premises. This may be held to be the case where no action is taken to stop them over a period of time, and this will often be the situation in the case of children.

Occupiers' Liability Act 1957

Definition of terms

Under *OLA 1957* the law imposes a duty on occupiers of premises in respect of lawful visitors. The definition of an occupier is based on control not ownership. Therefore a local authority could be liable for a person being injured while accessing a derelict building requisitioned by the council, but not yet actually purchased from the owner, the situation which occurred in *Harris v Birkenhead Corporation (1976)*. This would mean that a building contractor, who takes control over part or whole of a client's premises while undertaking construction work would effectively be the occupier. It is also possible for two or more parties to be occupiers under the act, and this would be the case where a landlord is responsible for maintaining the overall fabric of the building but the tenant has responsibility for internal repairs, the position in *Wheat v Lacon (1966)*.

The definition of the term 'premises' is very wide and goes well beyond the typical idea of a building or land, to include a vehicle as in *Bradford v Robinson Rentals (1967)*, a lift, as in *Haseldine v Daw (1941)*, or even a ladder, as in *Wheeler v Copas (1981)*.

The duty of care

Under *OLA 1957* the duty of care imposed on the occupier is the common duty of care set out in Section 2(2) of the act, which is defined as; 'to take such care as in all the circumstances of the case is reasonable to see that the visitor will be reasonably safe in using the premises for the purpose for which he is invited or permitted by the occupier to be there'.

This duty of care is effectively the same as the duty in common law Negligence, and the extent of the duty will depend on the nature of the claimant. The duty will be wider in the case of children, as per Section 2(3)(a) of the act, especially if the premises are an attraction to children, as in the case of a disused boat in *Jolley v Sutton L.B.C. (2000)*. Similarly, as per Section 2(3)(b) of the act, the duty will be seen as more limited, where the claimants are skilled employees who might be expected to take extra care. This was the case in *Roles v Nathan (1955)*, where two contractors installing a boiler were overcome by carbon monoxide gas while fitting the flue and the occupier was held not to be liable, as the contractors should have been aware of the danger.

The nature of the duty of care is not absolute, but is fault-based, so that the duty is limited by what is deemed to be 'reasonable' and this is decided by the court in the light of current attitudes. As a result, the occupier is not liable for obvious dangers, so in *Whyte* v *Redland Aggregates Ltd (1998)*, the occupier was not under a duty to warn people of the risk of drowning in flooded gravel pits, as such dangers should be obvious to most people. Similarly, in *Laverton* v *Kiapasha (2002)*, there was no duty on a shopkeeper to mop a wet floor when the shop was busy, as this was very difficult to do at the time and the hazard was obvious to the customers.

Defences

There are various defences available to the defendant, as set out under the act. So, in Section 2(4)(a), the defendant can rely on a reasonable warning notice in respect of adults, but he will have to take greater precautions in the case of children, such as locking up the premises to prevent access. It is also possible for an occupier to use an **exclusion or limitation clause** to restrict his liability, but such clauses are subject to the requirements of the *Unfair Contract Terms Act 1977(UCTA)*. Under the provisions of Section 2 of UCTA, an occupier cannot exclude liability for death or personal injury arising from his negligence, and any exclusion for losses to property must be 'reasonable'.

Another important defence is under Section 2(3)(b) of the act, where the occupier seeks to place the blame on a third party, such as a visitor or an independent contractor working on the site. The key determining factor here is the extent to which the occupier had any real control over the actions of the contractor. Where the work is exceptionally hazardous, the occupier would be expected to check if the contractor was competent, and he would be liable if he had not done this effectively. However, the occupier cannot be held liable for the negligence of highly skilled contractors, who appear to be competent. This was the case in *Haseldine* v *Daw (1941)*, where the occupiers had appointed what seemed to them to be a competent lift repair company to service a lift on their premises. When the lift failed suddenly, causing a fatal accident, it was the lift repairer who was liable, as the occupier could not be expected to exercise effective control over what was a specialist operation.

Finally, the occupier will not be liable if the claimant willingly puts himself in danger, this being the defence of 'volenti'. Such a defence is rare in the case of a lawful visitor, unless they act in an unexpected and foolish manner. However, the partial defence of contributory negligence is much more common, and this will apply where the actions of the claimant partly contributed to the accident. In such a situation, the claimant's award of damages will be reduced proportionally under the provisions of the *Law Reform (Contributory Negligence) Act 1945*.

Occupiers Liability Act 1984

Nature of the liability

Under common law, there was originally no liability to trespassers unless the occupier laid a deliberate trap. However, this changed in the 1970s with the development of the so-called common law 'duty of common humanity', whereby the courts held that child trespassers would be due a duty of care. So in *Pannett* v *Mc Guinness (1972)* the operator of an unfenced building site was liable for an injury caused to a young child by a bonfire, where rubbish was being burned but the operation was not being supervised.

The change in the common law was later followed by the passing of *OLA 1984*. This provides for limited protection in the case of trespassers, as well as for persons using a public or private right of way. The act creates a duty of care in respect of injury to the person due to the state of the premises, but does not allow for any claim for damage to property by the claimant. Under Section 1(3) of the act, the occupier owes such a duty if three circumstances apply. First, that the occupier was aware of the danger or had reasonable grounds to believe it existed; second, that the occupier had reasonable grounds to believe a person would come onto the property; and third, that it was reasonable in the circumstances to offer some protection.

In those cases where the danger is obvious, the occupier would be expected to be aware of the danger or he would be seen as negligent. However, he will not always be aware that a person is likely to come onto the property, unless there is evidence of trespassers such as graffiti or signs of forced entry. If there is no evidence that a person is likely to come into an area of danger, then no liability applies, as in *Higgs* v *W. H. Foster (2004)*, where the claimant fell into an inspection pit. As regards the reasonableness of the protection, this will be decided along similar lines as to what is 'reasonable practicable', balancing the risk as evident to the occupier against the cost in terms of time, money and effort to provide a means of protection.

Defences

As with the liability under the *OLA 1957*, there are a number of defences available under *OLA 1984*. Under Section 1(5) of the act this includes providing a suitable warning, although, as with *OLA 1957*, the suitability of the warning will be determined by the court, taking into consideration the age of the claimant and the possibility of the occupier taking further action such as securing the premises. In addition, under Section 1(6) the defence of 'volenti' will apply and the existence of an effective warning may help to prove that the claimant's actions fell into this category. So in *Ratcliffe* v *McConnell (1999)* the existence of a warning sign and the fact that the premises were locked provided clear evidence that the claimant's actions were totally due to his own volition. Therefore there was no liability on the defendants when the claimant was injured diving into a swimming pool, even though he was still able to enter the premises by climbing over the gate.

Defective Premises Act 1972

In addition to the liability under the *Occupiers Liability Acts 1957* and *1984* there is also a liability under the *Defective Premises Act 1972*, which may affect occupiers of premises. Under Section 4 of the act, where premises are let by a landlord who is under an obligation to maintain the building, he has a duty of care to all persons who come onto the premises, to ensure they are safe from personal injury. Although the act was designed to protect domestic tenants, it will also provide protection for a business leasing industrial premises, as well as a contractor who is injured working in the premises, whether domestic or not.

Product liability

Introduction

An employer has a liability not only for the actions of his employees, but also the hazards involved in the products he makes. In the case of non-consumers the liability lies in common law Negligence or breach of contract. The liability in Negligence is based on the so-called 'neighbour' principle, as established in *Donaghue* v *Stevenson (1932)*. As such, the employer is liable in respect of any person whom he can reasonably foresee as being liable to be affected by his acts or omissions. Therefore in the above case, the manufacturer was liable to a third party for a contaminated bottle of ginger beer, which led to the claimant suffering from gastroenteritis.

As regards an employer, the duty of care to ensure the safety of his employees would encompass not only injury caused by work equipment, but also any goods produced for sale, such as hazardous chemicals. We will also look at this in relation to contractual liability in the next chapter.

The Consumer Protection Act 1987

In the case of consumers, the liability in common law Negligence has been extended by a stricter duty of care under the *Consumer Protection Act 1987*, implementing the EU Product Liability Directive (85/374/EEC). The term 'consumer' means a person not acting in the course of business, who obtains goods normally used by consumers. More importantly, it covers not only such persons purchasing goods under a contract of sale, but also private individuals who have such goods purchased for them under a consumer contract entered into by another party.

Under Part 1 of the *Consumer Protection Act 1987* the law imposes a strict liability on producers, so any consumer no longer has to prove negligence. In order to claim the consumer needs to prove that the product is defective, in other words that it is not to a standard that could reasonably be expected. In determining this the court will take into consideration the description of the product, how it is labelled and marked, and any instructions provided with it. Liability here will lie with any producer or manufacturer in the EU, but it will also include any party importing the goods into the EU.

Although liability is strict, it is not absolute, and there are various defences available to the producer of the goods. These include the fact that the defect did not exist at the time of supply and that the technical knowledge available at the time meant that the producer could not have known about such defects.

So whether goods are produced for use at work or for use at home, the producer will have a liability in Negligence and under breach of contract to the workplace purchaser; and under the *Consumer Protection Act 1987* as well, in respect of any consumer sale. There is more on this topic in Chapter 4, which deals with contractual liability.

Conclusion

The civil liability of the employer for his employees is intrinsically linked to the criminal liability of employers, as often the civil action arises out of the same set of circumstances for which criminal liability is or may be imposed. In this respect, it is not surprising that the Tort of Negligence and breach of statutory duty are so closely interrelated and have developed in step with the Criminal Law.

However, with the ending of the right to take an action for breach of statutory duty, one of the key links has now gone, and the subordinate legislation only provides for criminal liability. In the common law action for Negligence the claimant has to prove fault on the part of the employer, which may be a little more difficult than relying on the stricter liability sometimes imposed by statute. However, we have seen that the court can reverse the burden of proof through the principle of 'res ipsa loquitur', while the use of a criminal prosecution as evidence will often be conclusive to the civil case, given that the standard of proof required in a civil action is lower.

We can see that the area of occupier's liability poses problems for employers, both as occupiers of premises as well as landlords of premises used by other employers. The interrelation of statutory and common law liability has created a complex area of law, which needs to be taken into consideration in respect of civil liability. In addition, the occupier should be aware that there is also a concomitant criminal liability for employers under Section 4 of the *Health and Safety at Work Act 1974*. In contrast, product liability tends to be more limited in its impact on health and safety, but we should still be aware that it can be of importance, especially if an employee is injured by defective goods or articles used at work. We should also note here the corresponding criminal liability under Section 6 of *HSWA*.

The recent trend to place liability on the employer in civil cases wherever possible might now be coming to an end, influenced by increasing concerns over the development of the so-called 'litigation culture'. However, it is possible that the courts may eventually move towards a narrower inter-pretation of what is 'reasonably foreseeable', so as to impose a stricter liability and to deal with situations where fault is difficult to find, but there is a perceived need to provide a remedy for an injured claimant. However, we will have to see how legal developments work out in this area, as the law is clearly still in a state of flux.

Notes

1   Nigel Tomkins 'First principles in employer's liability' *Journal of Personal Injury Law* 2010, 131–138
2   Maria Lee 'Tort law and personal injury in the regulatory state' *Journal of Personal Injury Law* 2011, 3, 137–143; and also Frances McCarthy 'Case comment: Barker v Quantum Clothing Group Ltd' *Journal of Personal Injury Law* 2011, 3, C122–127
3   Stephen O'Doherty 'Personal injury: causation: a floating concept' *New Law Journal* 2009, 159, 809
4   See David Branson 'Tempting fate' *Safety and Health Practitioner*, Jan 2010; and also Howes V. 'Who is responsible for health and safety of temporary workers? EU and UK perspectives' *European Labour Law Journal* 2011, 2, 379–400

4 Contractual liability and the Tort of Nuisance

Introduction

Although liability in Negligence is now the basis of civil liability for health and safety as regards accidents at work, contractual liability still plays an important part in other aspects of health and safety. In this chapter, I would like to look at the main types of contractual liability and their impact in this area. I will be looking at the principles of Contract Law, before looking at the employment contract and then the statutory regulation of business contracts under the *Sale of Goods Act 1979 (as amended)* and the *Supply of Goods and Services Act 1982*. In addition, I will say a little about liability in the Tort of Nuisance, and especially the rule in the case of *Rylands v Fletcher (1868)*.

Contract Law

The nature of contractual liability

Contracts provide the legal underpinning for most business agreements, including the sale of goods or supply of services, as well as the creation of employment arrangements. While business contracts are often made in writing, they can be made verbally or by way of a deed. A contract is essentially an agreement between two or more parties, which creates rights but also imposes obligations. Such an agreement may be enforced by the courts, but it is also the case that it might not, so it is essential to know what are the requirements of a valid contract, as well as which factors may make it unenforceable or invalid.

Forming a valid contract

To constitute a valid contract, it is necessary to have a valid offer made by one party (the offeror) and an acceptance made by another (the offeree). Offers and acceptances must be unambiguous and intended as such for them to be valid. In addition, the offeree must usually give something of value to the offeror in exchange for the subject of the contract, unless the contract is made by deed. Moreover, it must be clear that the parties intend to be legally bound, although this will be presumed in business agreements, including contracts of employment.

The terms of the contract may be express or implied. In the former case they are clearly stated verbally or written into the contract. Alternatively, terms can be implied by statute or by the courts in common law. Statutes such as the *Employment Rights Act 1996*, the *Sale of Goods Act 1979* and the *Supply of Goods and Service Act 1982* all imply terms into the contract, as detailed below.

In addition, the law may restrict the use of contract terms, especially those excluding or limiting liability, as under the *Unfair Contract Terms Act 1977*. Moreover, onerous terms may be struck down by the courts either under the above legislation, or on the basis of being unreasonable or unconscionable. An example here is the case of *Schroeder* v *Macauley (1974)*, where a contract between a recording company and a composer was struck down because the recording company tried to impose terms that were seen as very one-sided.

Invalidating factors

A contract may be invalidated for a number of reasons. First, the parties may not have the **legal capacity** to contract in the first place. Many contracts can only be enforced against adults, so persons under 18 will need an adult guarantor. A contract may be invalidated by one party being mistaken as to the nature of the agreement, because it will be argued that there was no genuine agreement between the parties. Similarly, if there is a **misrepresentation** by one party, then the contract may also be invalidated, as will be the case if one party exercises **undue influence** or duress to compel the other to enter into the contract, as all of the above puts in question whether the agreement was genuine in the first place.

Remedies for breach

If a contract is breached, the main remedy is damages or payment of monetary compensation. The aim of the damages is to put the injured party in the same position as if the contract had been performed. This should be compared to the position in tort, where the purpose of damages is to put the parties into the position they would have been in had the tort not occurred in the first place. The key difference is that the injured party can claim for any profits he could have made had the contract been performed. This could be an advantage in certain cases involving an accident at work. For example, if a self-employed person is injured due to the negligence of the person engaging him, or any other party he has a contract with such as a subcontractor, he could claim for any loss of profits due to being unable to complete the work. In addition, he could claim for loss of future profits for the period during which he is incapacitated.

The other forms of remedy are specific performance, requiring a party to perform the contract, or an injunction preventing him from pursuing a course of action. In the case of causing death or injury to a person at work, this is not particularly useful. However, if we are talking about the supply of defective equipment, the remedy of specific performance could be used to obtain a replacement machine, which may be more useful than payment of damages, especially if it is difficult to obtain such equipment elsewhere.

Contracts of employment

Definition of employment

The most important contract we need to consider in respect of health and safety is the contract of employment. We must distinguish here between a contract of service, which is a contract of employment, and a contract for services, which is a contract of self-employment. This is very important, because statute law provides a wide range of protection for employees under a contract of employment, including rights to claim unfair dismissal, redundancy and limits on working hours. None of these protections exist for self-employed contractors, so it is not difficult to see why employers have sought to engage workers on a self-employed basis, rather than as employees. In addition, there are tax advantages to the worker in being self-employed, so both parties may have a reason to define their contract as self-employment.

The leading case that defined the difference between employment and self-employment was *Ready Mixed Concrete* v *Ministry of Pensions (1968)*. This case established that the courts could determine the true nature of the relationship, independent of what is stated in the written contract. In order to do this, the courts will use a number of tests to determine the nature of the employment relationship. The first of these is the so-called 'control test', whereby a person is seen as an employee if the person he works for has effective control over his actions. However, this test falls down in the case of professional or skilled employees such as doctors or engineers, as the employer will often have little real control over how they do the job. A second test is the 'integration test', whereby the court will look at the extent to which the worker is integrated into the organisation, sharing in pension arrangements and training schemes. However, the changing nature of employment has made it very difficult to distinguish between persons employed on a regular basis and persons who are 'self- employed' but who work full time for a specific organisation. The final test is the so-called 'mixed test', which looks at all of the above criteria and was the test actually used in the *Ready Mixed Concrete* case.

In respect of health and safety matters, the courts have tended to see the worker as an employee in most cases, even when the contract states that they are self-employed, as was the case in *Lane* v *Shire Roofing (Oxford) Ltd (1995)*. In the case of workers supplied by employment agencies, the agency is now seen as the employer under the provisions of the *Agency Worker Regulations 2010*. A worker is only likely to be seen as genuinely self-employed for health and safety purposes if he provides his own tools and is clearly operating a business on his own account. In addition, it would be expected that such a person would have taken out some form of personal indemnity insurance, as well as providing for their own insurance against personal injury.

Contracts of employment

In a contract of employment the nature of the legal agreement is substantially affected by statutory controls, in particular the *Employment Rights Act 1996*, the *Working Time Regulations 1998* and the *Equality Act 2010*. All of these pieces of legislation impact on health and safety in the employment relationship, but in addition the courts imply in terms from common law, which also have an effect. The main one is the duty of care on the employer to look after the health and safety of his employees, a duty of care that is effectively the same as the one that provides the basis of liability in common law Negligence. While the relevant statutes lay down the basic guidelines, it is decided cases that are used to clarify the specific nature of the legal liability.

Terms of the contract

Under Part 1 of the *Employment Rights Act 1996 (ERA)*, the employer has to provide a written statement of the particulars of employment, which includes such matters as the names of the employer and employee, the date when employment began and when continuous employment is deemed to begin, as well as the hours of work and holiday entitlement. While the contract will normally be in writing, if no written contract is provided, the courts can determine the terms and conditions according to what they believe is reasonable, based on the minimum statutory requirements. Any changes to the contract terms must be agreed by both parties.

Hours of work

The key health and safety issues relate to the hours of work and holiday entitlement. Hours of work have long been regulated in the case of specialised employment categories, such as lorry drivers (see EU Regulation 3820/85). However, with the introduction of the Working Time Directive (93/104/EC), restrictions on the hours of work have now been extended to cover the category of 'workers' as a whole. This is a wider term than 'employees' and will cover casual employees and self-employed workers who are not operating a business on their own account.

The maximum weekly working hours are not to exceed 48 hours, averaged over a 17-week period. However, it is possible for the employee to opt out of the legal control by way of a waiver, freely signed by both the employer and the employee. There are also restrictions on the number of hours that can be worked by employees under 18 and the hours of work for night workers. In addition there are minimum rest periods required of 11 hours in every 24 hours, and rest breaks of 20 minutes in any 6 hour shift, although the latter may be varied by a workforce or collective agreement. An employee is entitled to an annual paid holiday entitlement of 28 days each year (including Bank Holidays).

Clearly the limitation on working hours and entitlement to holidays and break-time is a health and safety issue. Long hours of working lead to fatigue and this is one of the main causes of accidents at work. It was for this reason that the Working Time Directive was introduced under Section 153 (previously 137) of the *Treaty of Rome* as a health and safety matter. Consequently, the 'opt out' arrangements that exist in the UK are cause for concern. Although they are supposedly based on the free consent of both parties, there is evidence that prospective employees often feel that have no option but to agree to the 'opt out', in order to secure employment.

Another problem is that it can be very difficult to distinguish between health and safety and non health and safety matters in the contract of employment. So the issue of minimum pay or **zero hour contracts** has an impact on health and safety, as persons may be forced to work excessive hours to earn enough to live on. However, such issues are not always seen as health and safety matters, and that is important because under the *ERA* there is extra legal protection for employees raising health and safety issues at work, particularly if this leads to dismissal. If the court finds that the issue is not a health and safety matter, then that protection can be lost. This can be seen in the case of *Kerr v Nathan's Wastesavers Ltd (1995)*, where the claimant was dismissed when he refused to drive a vehicle that he believed was overloaded. The Employment Appeal Tribunal held that there was no real health and safety issue involved and so there was no right to claim unfair dismissal as he had failed to obey a reasonable order from his employer.

Health and safety issues

The relationship between employers and employees will be based on a contract of employment. This should be in writing as required by the *ERA*, but even if there is no written contract, a contract can be construed by the courts if necessary. In all such contracts there is an implied term that the employer has a duty of care to his employees to ensure their health and safety, in effect a similar requirement to the duty of care in common law Negligence. Moreover, such a contract term cannot be excluded or limited by an express clause, by virtue of the *Unfair Contract Terms Act 1977*, which applies to all employment contracts in favour of the employee, a point made clear in the leading case of *Johnstone v Bloomsbury Area Health Authority (1991)*.

In the *Johnson* case, the Court of Appeal held that the employer was in breach of his duty of care, because he required the employee to work up to 88 hours a week, as a result of which the employee began to suffer from depression. The court held that such long hours were a threat to the health and safety of the employee. This was later confirmed by the introduction of the Working Time Directive (94/104/EC), which made it clear that working hours are a health and safety issue, as the directive was passed under the health and safety requirements of the Treaty of Rome.

Although a duty of care to ensure the health and safety of the employee is an implied term in the employment contract, the Tort of Negligence has become the main means of seeking redress in the UK. However, the contractual remedy may still be of use if the injury occurs abroad, as an action for Negligence is limited to the United Kingdom. The key issue here is whether the employee has a contract of employment deemed to be regulated by UK law, in which case an action can be brought for breach of contract in respect of the implied duty of care. This was the case in *Matthews v Kuwait Bechtel Corporation (1959)*, a case brought by an employee injured while working abroad.

While an action for breach of contract is always possible, it should also be noted that there is a problem if the action is for latent injury, such as an industrial disease like asbestosis. Although the limitation period in contract is six years, it runs from the time of the injury occurring, which is when, for example, the asbestos fibre is inhaled. Given the length of time for such a disease to develop any action will certainly be out of time. Under the provisions of the *Limitation Act 1980* the limitation period in Tort of Negligence runs from the date when the disease is discovered, which is much later.

Unfair dismissal

Under Part X of the *ERA*, which consolidated earlier employment protection legislation, employees have the right to claim for unfair dismissal at an Employment Tribunal, but they must have worked for at least one year or two years if appointed after April 2012. However, if the matter involves a health and safety issue, then there is no qualifying time period and the action is brought under Section 100 of ERA. This allows for an employee to claim for unfair dismissal, where he is dismissed after bringing a health and safety matter to the attention of his employer, such as refusing to work in a place of danger or taking other reasonable steps to protect himself in such a situation. An example here is *British Aircraft Corporation v Austin (1978)*, where the employee was dismissed after refusing to work, when the employer failed to supply him with safety glasses.

In a case of unfair dismissal, the employee can seek various remedies: to be reinstated to his old job, re-engaged by the employer in another job, or he can claim compensation payments, which can be quite high as there is a special award for persons dismissed on health and safety grounds.

In addition to straightforward dismissal, there are remedies in the case of a person being selected for redundancy, when the effective reason is that the employer wants to dismiss the person for raising a health and safety issue. The main problem is being able to prove that the redundancy was not the result of genuine economic reasons.

Under Section 44 of ERA, an employee can also claim if he suffers a detriment as a result of a health and safety matter, such as being disciplined or having a reduction in wages. Again, the action is brought to an Employment Tribunal, which can award compensation to the injured party.

When taking a claim under Section 100 or Section 44 it is important to show that the matter was a health and safety issue, otherwise the matter becomes a misconduct issue and the employee cannot obtain protection for disobeying a reasonable order. In *Buttars* v *Holo-Krome Ltd (1979)* the employee refused to operate a machine claiming that it was inadequately guarded, despite the Factories Inspector stating that this was not the case. The tribunal held that this was an example of misconduct and the employee could not make a successful claim for unfair dismissal.

Discrimination at work

The key legislation now is the *Equality Act 2010*, which covers discrimination on the grounds of age, disability, pregnancy, race, religion, sex and sexual orientation. The new act replaces a variety of older legislation, such as the *Sex Discrimination Acts 1976* and *1986*, the *Race Relations Act 1976* and the *Disability Discrimination Act 1995*. The employee can take an action to the Employment Tribunal, who can order the employer to take steps to remedy the situation or will award compensation in lieu, where it is possible to show discrimination. Discrimination here can be direct or indirect, in the latter case this involves setting out criteria for appointment, which make it difficult or impossible for a person of a specific type to obtain employment or carry out a particular role.

As regards health and safety, there may sometimes be a conflict between the requirements of this legislation and the need to ensure a safe workplace, in which case health and safety takes priority. So in the case of a disabled person, there is a requirement that the employer makes 'reasonable adjustments' to make sure that the disabled person is able to carry out the job. This can involve installing a ramp to access a building, modifying the workstation or installing a visual fire alarm for deaf people. However, the employer should not put the safety of the employee or other employees at risk by engaging someone whose mental or physical disability makes it dangerous for them to do the job. An example here would be *Singh-Deu* v *Chloride Metals Ltd (1976)*, where a person suffering from a mental illness was dismissed because his job involved handling hazardous chemicals.

In the case of sex discrimination, a similar conflict of interest may arise. On the one hand a large number of restrictions have been lifted, such as on women working in mines, for example. However, a woman may be unable to carry out heavy lifting work so would not be able to be employed on certain tasks. Similarly, they may not be able to work with some chemicals if they are of child-bearing age. A leading case here is *Page* v *Freight Hire (Tank Haulage) Ltd (1981)*, where a female tanker driver was dismissed because the chemical she was transporting was deemed be harmful to unborn children and the employer was unable to offer her alternative work. The dismissal was upheld as reasonable by the Employment Tribunal, even though the claimant argued that she did not intend to have children.

Discrimination on the grounds of race or religious belief may also cause some problems. The unwillingness of Sikhs to remove their turbans for religious reasons led to a compromise on the wearing of safety helmets. Under the provisions of *Section 11* of the *Employment Act 1989* Sikhs working on building sites do not have to wear safety helmets, but their ability to claim compensation in the case of injury is limited. However, where religious requirements may impact of the safety of others, then such a derogation from the law would not be allowed. So in *Panesar* v *Nestle & Co. Ltd (1980)* a Sikh was lawfully refused employment at a food processing plant, because there was a company policy of requiring employees to be clean shaven to prevent contamination of the foodstuffs.

In this way we can see that issues of health and safety are linked with issues of employment and discrimination. The courts will have to try and balance the competing interests of employers and employees in this area, and this is of course an area of continuing change. The recent dispute over the wearing of crucifixes at work by some Christians, or the wearing of the hijab by some Muslims, is just another example of the problems faced in this area.

Whistle blowing

When employees are concerned by a breach of the law relating to health and safety, they may feel that they cannot report such an action, because they will be putting their job in jeopardy. As a result, legal protection has now been offered to employees under the *Public Interest Disclosure Act* 1998 *(PIDA)*, the so-called 'whistle blowing' legislation. This was passed in the wake of a series of scandals, in which employees who brought misdeeds to light ended up being dismissed by their employers, and, in some cases, subject to legal proceedings for breach of confidentiality.

The act allows employees to disclose certain information, known as a 'qualifying disclosure' to specified organisations of the state, such as the HSE and local authorities and even to the media. Such information is defined as including information, which the employee reasonably believes, to involve the carrying out of a criminal offence or matters that may endanger the health and safety of an individual or cause environmental harm.

This legislation is of crucial importance given the increasing tendency for employers to insert confidentiality clauses into their contracts of employment, seeking to prevent employees from revealing information about the organisation that they do not wish to see published. In addition, employees who are made redundant or who are dismissed are often required to sign a document preventing them from revealing information as part of a contractual settlement. The danger is that such actions will tend to prevent important information from coming into the public domain, which may concern wrongdoing by the employer.

As a result, the act allows the making of such qualifying disclosures directly to organs of the state such as a government minister, the HSE or a local authority. Also, a disclosure may be made to non state organisations, such as a charity or even the press, so long as the disclosure is made in '**good faith**', not for personal gain and where the employee reasonably believes he will suffer a detriment if it is disclosed to the employer. If an employee is dismissed following the making of such a disclosure, then it is automatically unfair and an employee can claim unfair dismissal under the provisions of Section 43 of the *ERA 1996*. As such, the employee can claim for reinstatement, re-engagement or compensation, and in this case the maximum compensation of £50,000 for unfair dismissal does not apply and may be exceeded.

The effect of *PIDA* on health and safety matters will take time to assess. However, it does provide some protection for employees who report health and safety issues to the inspectorate. However, it is notoriously difficult to prove that an employer is dismissed for a particular action at work, especially if the employer resorts to making redundancies at a later stage. It also does not address the situation of an employee who signs a so-called 'gagging agreement', as the employer may be able to stop any pension payable.

Contracts for goods and services

Introduction

If we wish to look at business contracts involving health and safety issues, we need to look at contracts for the sale of goods as well as contracts for the provision of services. Both are covered by different legislation, namely, the *Sale of Goods Act 1979 (SOGA)*, as amended by the *Sale and Supply of Goods Act 1994 (SSGA)*, as well as the *Supply of Goods and Services Act 1982*, both of which overlap to some extent. In addition, we need to consider the effect of Section 6 of the *HSWA*.

Sale of goods

The *SOGA 1979* (as amended by *SSGA 1994*) implies terms into a contract of sale, which provides protection for the purchaser. The protection here differs between a consumer and a non-consumer contract. A consumer contract is one in which the seller acts in the course of business and the purchaser does not, while a non-consumer contract is one is where both parties act in the course of business. For the most part, we will be concerned with non-consumer sales, but the position of consumer sales will also be considered.

In all transactions, *SOGA* implies in a duty under Section 13 of the act, that the goods sold must correspond to the description given to them. In Section 14 there is an implied term that the goods must be of 'satisfactory quality' or fit for purpose. In this respect 'satisfactory quality' will include factors such as safety and durability. Section 15 requires that where goods are sold by sample, they must correspond to the sample.

Non-consumer contracts

In a non-consumer contract, it is possible for a seller to exclude or limit liability under Sections 13 to 15 of *SOGA*. However, under the provisions of the *Unfair Contract Terms Act 1977*, it is not possible to exclude liability for death or personal injury, and any other exclusion will only be allowed if the court feels it is reasonable. Nevertheless, this will enable the parties to such a contract to determine the liability each party wishes to incur, so that they can then arrange appropriate insurance cover. The relevant costs can then be built into the contract price.

With regard to health and safety matters, the main issues involve the sale of goods that are used by an employee in the course of his employment, such as any work equipment or substances for use at work. Any equipment supplied must comply with the requirements of the *Supply of Machinery (Safety) Regulations 2008*, as well as the *Provision and Use of Work Equipment Regulations 1998* and Section 6 of the *HSWA*. If work equipment does not comply with the above, then it is up to the seller to make the purchaser aware of this, so that the relevant alterations can be made to make the equipment compliant.

If anyone is injured when using the equipment, the employer or person in control is primarily responsible under common law Negligence. However, the purchaser of the machine can then join the supplier as a co-defendant in any civil action arising. Moreover, he can sue the purchaser for breach of contract under *SOGA*, on the basis of a breach of Section 14, as the goods are probably not of satisfactory quality or fit for purpose. If the failure of the equipment leads only to physical damage to the workplace, then the liability may be subject to any reasonable exclusions or limitations in the contract. In such an action the purchaser will be able to claim for any financial losses arising from the breach, such as increased insurance premiums and loss of goodwill.

In the case of any substance supplied it has to be safe as per the requirements of Section 6 of *HSWA*. If the substance is not safe for use at work in accordance with *HSWA*, then it is clear that it is not of satisfactory quality. As with defective equipment, the purchaser can sue the supplier for breach of contract in respect of the implied terms under Section 14 of *SOGA*, subject to any 'reasonable' exclusion or limitation clause, although again such a clause will not apply if death or personal injury arises as a result of the breach.

Consumer contracts

In the case of a consumer contract, it is not possible for the seller to exclude or limit liability for Sections 13 to 15 of *SOGA*. This means that a supplier of goods will be liable for any financial loss caused to that person as a result of any accident arising from the use of faulty or defective goods, including loss of earnings. However, if the goods are purchased by a self-employed contractor for use both at home and at work, it might be argued that a 'reasonable' exclusion clause could apply in respect of any loss of earnings, resulting from any accident caused by the use of the goods at work, as this may be seen as being a non-consumer contract.

Supply of goods and services

Introduction

Where goods are supplied as part of a contract for services, such as supplying vehicle parts when repairing a piece of machinery, then the *Supply of Goods and Services Act 1982* will apply, as is the case where the contract is one purely for services. Any goods supplied under the contract are covered by exactly the same legal regime as goods sold under a simple contract of sale, so that goods that are unsafe will be in breach of Section 4 of *SGSA 1982*, which requires goods to be of satisfactory quality and fit for purpose. As regards the supply of any service such as the inspection of lifting equipment, under Section 13 of *SGSA*, the service has to be done with 'reasonable care and skill', a standard of care analogous to the common law duty of care in Negligence. As a result, there is no real advantage in suing for breach of *SGSA* in respect of any failure to perform a service properly, except where the work is done outside the UK, when liability in tort will not apply.

Leasing out of equipment

When equipment is leased out by one party to another, the party leasing the equipment (the lessor) will retain the duty to ensure that the equipment remains in good working order. Indeed, in such a situation, the lessor may well have been held liable under any relevant statutory duty, until the recent abolition of the civil claim for breach of statutory duty. Therefore, in the case of *Ball v Street (2004)* the lessor was seen as falling within the requirement of Section 3(3)(b) of *PUWER 1998* in respect of a piece of agricultural equipment leased to a neighbouring farmer.[1] This is because the lessor was seen as retaining control over the equipment. The liability now would lie in common law Negligence, with the lessor owing a duty of care to the user of the equipment.

When goods are provided to another party under a hire purchase arrangement, the obligation of maintenance will be seen as passing to the person it is hired to, even though in law, ownership of the goods remains with the person who is hiring out the equipment. However, the supplier of the goods will retain liability for the condition of the goods when supplied and for a reasonable time afterwards, in accordance with the expected period of 'durability' needed to satisfy the requirement of 'satisfactory quality'.

The Tort of Nuisance

Introduction

The law relating to Nuisance is only partly related to health and safety matters, but it is of greater relevance when we look at liability for the storage of dangerous materials on the land. In this respect, it is analogous to the liability under the *Occupiers Liability Acts 1957* and *1984*, except that liability is to persons who are not on the land itself but on land nearby. In particular we need to focus on the case of *Rylands* v *Fletcher (1868)* and the more recent case of *Cambridge Water Co.* v *Eastern Counties Leather (1994)*, which profoundly modifies the effect of *Rylands* v *Fletcher*. Most of the liability here is based on common law, but there are various statutes that impose liability, such as the *Environmental Protection Act 1990*, which we will consider briefly later.

The Tort of Nuisance was originally land based, but in recent times it has developed into a wider liability with the growth of Public Nuisance and Statutory Nuisance. While Private Nuisance is still related to the ownership of land and provides a remedy for interference with that land, Public and Statutory Nuisance are not related to land ownership, but provide a remedy for a wider range of potential claimants. I will look at both of these areas of law below.

Private Nuisance

Nature of the liability

The Tort of Nuisance has been defined as 'an unreasonable interference for a substantial length of time, by an owner of property, with the use or enjoyment of a neighbouring property', as in the case of *Cunard* v *Antifyre (1933)*. This means that the claimant must have an interest in the land affected, which may involve ownership or the holding of a tenancy, but it will not extend to another person who happens to live on the land. This was made clear in the case of *Hunter* v *Canary Wharf Ltd (1997)*, which overruled an earlier decision to that effect in *Khorasandjian* v *Bush (1993)*.

In order to claim compensation, the claimant needs to show evidence of unlawful interference with the use of his land. This will include such matters as allowing smoke, noise, noxious materials or toxic substances to come onto the claimant's property. Such interference has to be of a continuous nature, but it can include a single escape of material onto the claimant's land if there is evidence that the same problem has happened before. As a result, in the case of *British Celanese* v *Hunt (1969)* the court held that the escape of tin foil onto the claimant's property was a Private Nuisance because the foil had blown onto the premises before.

The more recent case of *Crown River Cruise* v *Kimbolton Fireworks Ltd & Others (1996)* even allowed a claim for Private Nuisance in respect of a 20 minute firework display, which showered the claimant's barge with flammable material, causing it to set alight. It would seem that the difference between common law Nuisance and common law Negligence is rather limited here, and the claimant may be able to choose which action to take. There may be a benefit to taking the action in Nuisance, as the rules on claiming for economic loss or loss of profits are more favourable.

Unreasonable interference

It should be noted that the claimant needs to show that the interference here was unreasonable. In determining what is unreasonable, the court will take into consideration various factors. The first one is the conduct of the defendant in the circumstances. This is of importance where the reason for the interference with the claimant's land is due to factors beyond his control. So in *Leakey* v *National Trust (1980)* the escape of material onto the claimant's land was due to natural causes and, as such, the liability was negated because it was beyond the means of the defendant to deal with it. A similar decision was reached more recently in *Arscott* v *The Coal Authority (2004)*, where the defendant was not liable for the effect of flood water being diverted from his property to the claimant's, as the floods were seen as a 'common enemy' and not a nuisance caused by the defendant.

In addition, the courts will take into consideration factors such as the locality of the property where the nuisance occurs. As such, the court may accept that in an industrial area the claimant should be expected to deal with different activities than those which would be encountered in a predominantly residential area. In the case of *Gillingham Borough Council* v *Medway (Chatham) Docks Ltd (1993)* the claimant sued for the effect of noise due to the increased vehicular traffic going to the dock, but the claim was rejected, as it was held that this was an industrial area and such activities had to be accepted as part of everyday life. It is also to be noted that the character of a locality can change over time and this can mean that interference that was seen as reasonable may eventually become unreasonable, or vice versa.

The courts will take into consideration the sensitivity of the claimant and may find that what he claims to be a nuisance would not normally be seen as a nuisance to others. A recent case here is *Network Rail Infrastructure Ltd v Morris (t/a Soundstar Studio)(2004)*, where the claimant's recording studio was unable to prove Nuisance in respect of the use of ultrasonic equipment, as this was now in general use and the general public were expected to cope with it. However, it would appear that there is still a liability if there is evidence that the nuisance would cause damage to the claimant's property, as made clear in the earlier case of *McKinnon Industries Ltd v Walker (1951)*.

The court will also look at the public utility of the defendant's activities, and balance them against the interests of the claimant. This may be of help where the defendant is providing employment or operating an industry deemed to be vital for the country as a whole. This will mean that noise from construction activity will have to be accepted, at least during normal waking hours, while the project is being completed. More controversially, the noise of aircraft is also something that the court may find as an acceptable consequence of a necessary activity, as in the case of *Dennis v Ministry of Defence (2003)*.

Finally, it is important to determine whether there is an alternative statutory remedy that may operate here. So in the case of *Marcic v Thames Water Utilities (2003)* the claimant's house was affected by foul water that had come from the defendant's sewerage system. The court held that there was an alternative remedy under the provisions of the *Water Industry Act 1991*, which took into consideration the interests of the individual affected and the community as a whole.

Remedies for Private Nuisance

Once the above issues have been weighed up by the court, they can decide if there is an issue of Private Nuisance. If this is the case, then there are two main forms of remedy; these are an injunction or an award of damages.

An injunction is an order by the court to desist from an activity, such as allowing toxic material to come onto the claimant's property. This is an **equitable remedy** and as such will only be awarded by the court if it thinks it is appropriate. In this respect, they will take into consideration the extent of the loss, whether it is possible to provide damages as an alternative and whether there is a public interest in denying an injunction.

Where an injunction is not awarded the claimant will get damages. This is a money payment designed to compensate him for any reasonably foreseeable losses caused by the nuisance. In this respect, the level of damages will reflect the extent to which the nuisance has caused the value of the property to depreciate.

Defences

The main defence to an action in Private Nuisance is that the defendant had been carrying on his operation for some time, without anyone seeking to take legal action to stop him. However, the defence is not total, and it is possible for the claimant to argue that the nature of the locality has changed in the meantime and the defendant's operations now constitute a nuisance. In addition, the defendant may be able to argue that his operations have been allowed for by statute, such as the right to construct an oil refinery, as in *Allen* v *Gulf Oil Refining Ltd (1981)*.

Public Nuisance

Nature of the liability

Unlike Private Nuisance, which mainly protects property interests, Public Nuisance covers a much wider range of issues including liability for personal injury, and the action is not limited to parties with a legal right in the affected property. The definition of a Public Nuisance is to be found in the case of *Attorney General* v *PYA Quarries Ltd (1957)*, which defines a public nuisance as 'one which materially affects the reasonable comfort and convenience of life of a class of Her Majesties subjects'. A 'class' of subjects here means any group of people who are similarly affected and such a group may include for example local neighbours on an **estate**, local businesses or users of a public highway.

Actions for breach of Public Nuisance are usually brought by members of the public against individuals or organisations whose actions cause them specific or particular damage, which means damage over and above that suffered by other members of the public. It can include such matters as damage to property caused by flying golf balls as in *Castle* v *St Augustine's Links (1922)*, or the failure to clear up pigeon droppings from a railway bridge, which were causing a danger to health as in *Wandsworth L.B.C.* v *Railtrack PLC (2001)*.

It is clear here that action must be taken by a group of persons and not just one person, as the wrong is effectively against society as a whole. It is for that reason that Public Nuisance will also constitute a criminal wrong. As such, there is a more direct link to health and safety issues, as a group of workers could use Public Nuisance to take action against an employer whose actions is causing them personal injury. More realistically, it will be brought by a group of residents against injury caused by the actions of such a defendant employer.

We have seen that such an action will usually be brought in Negligence, because Negligence actions are usually brought by individuals, not a group of persons, and because the Negligence action provides for better remedies in the case of latent injury. However, it is always possible for a group of workers to take an action for Public Nuisance as an alternative to an action for Negligence, if the nuisance causes an immediate injury. The main advantage here is that it is possible to claim for economic loss in Public Nuisance, in other words, loss of profits. While this is of limited value for employees, it may be of importance if the claimants are self-employed, as not only may their present earnings be adversely affected by a Public Nuisance, they may also lose future profit-making opportunities.

Remedies

As in Private Nuisance, the remedies include both an injunction and damages. As with Private Nuisance, a court will only award an injunction if it feels that it is justified in the circumstances. As an alternative, damages can be awarded, and they will cover compensation for personal injury, as well as loss of earnings or loss of profits as mentioned above.

Defences

The defences to Public Nuisance are as for Private Nuisance. These will include statutory authority, as in *Allen* v *Gulf Oil Refining (1981)*, mentioned above. In addition, the defendant may argue that he has obtained the necessary planning permission to carry out a particular activity. However, it should be noted that this does not constitute a complete defence to an action, as can be seen in the case of *Wheeler* v *Saunders (1995)*, in which the operation of a pig farm near to a residential development was found to be a nuisance, even though planning permission had been granted by the appropriate authorities.

The rule in Rylands v Fletcher (1868)

Introduction

The rule in Rylands v Fletcher was a response to growing industrialisation, which posed a threat to the property of other landowners because of the escape of hazardous substances onto their land. The result was the creation of a liability in tort, which imposed strict liability for the escape of such substances where they caused damage to neighbouring land. As such, it is essentially a land based tort like Private Nuisance, but it has a significant impact on the safe operation of industrial premises.

The case of *Rylands* v *Fletcher (1868)* involved an escape of water from a reservoir constructed by the defendants. The reservoir was built on the top of the entrance to several disused mines and the defendant failed to ensure that these entrances were effectively sealed so that water could not escape from the reservoir. As a result, water flooded through the mineshafts and into working mines on the claimant's property, causing substantial damage.

Nowadays the action would probably be brought in Negligence, but at the time, liability in Negligence was not sufficiently developed. Instead, the court developed a new legal action to impose strict liability in the event of an escape of hazardous substances onto neighbouring land.

Requirements of the tort

In order to establish liability it was essential to prove various elements of the case. First, the defendant had to bring something onto his land that was not natural to the land. This would not include natural vegetation, but would include new crops planted there, as well as accumulations of waste materials or other hazardous substances. Second, the material had to be brought onto the land for the purposes of the defendant, and the use of the land had to be non-natural.

This concept of non-natural use became the basis of the liability, distinguishing it from other types of Nuisance. However, the concept of what is non–natural is very fluid and has changed over time. In *Mason* v *Levy Auto Parts Ltd (1967)* it was held that the storage of flammable materials on the defendant's land constituted a non-natural use. Consequently, he was liable for damages under the tort of Rylands v Fletcher, when some of these materials ignited and the fire spread to a neighbouring property causing damage. However, later cases have considerably limited the definition of what is non-natural, so that in *Transco PLC* v *Stockport Metropolitan B.C. (2004)* it was even held that the transport of water though underground pipes did not constitute a non-natural use of the land. It would seem that the case of Rylands v Fletcher itself, if brought today, might actually fall outside the rules of the tort that bears its name.

Finally, it is important that the substance actually escapes from the defendant's land or there is no liability under the tort. So in *Read* v *Lyons Ltd (1947)* the claimant was injured in an explosion at a munitions factory during the war. However, as there was no escape of material from the premises, there was no liability under the rule in Rylands v Fletcher. Instead, an action would only lie in Negligence, and this was difficult to prove in this case. Nowadays, it is much more likely that the defendants would have been liable in Negligence and so there would have been no need to pursue a strict liability action.

The nature of the liability

The action under Ryland v Fletcher is a development of the Tort of Private Nuisance, so it could be argued that the action is limited to persons with an interest in land in the area of the release, which would effectively limit its use as a means of claiming for injury at work. However, several cases have been brought by persons with no proprietary interest in the land, in particular the case of *Rigby* v *Chief Constable of Northamptonshire (1985)*, involving an escape of CS gas. In this case, it was held that an action under Rylands v Fletcher would apply in respect of an escape of a hazardous substance onto the public highway, which caused injury to the public. More recently it has been suggested that a better remedy lies by suing for breach of the *Human Rights Act 1998*, as in the case of *McKenna & Others v British Aluminium Ltd (2002)*, where residents were affected by pollution from a factory. However, following on from the decision in *Hunter v Canary Wharf (1997)* it would appear that a right of action in Nuisance may increasingly be limited to parties with a proprietary interest in neighbouring land.

The type of damages that a person could recover damages for in Nuisance was originally quite wide, but has recently been limited by decided cases. A claim for personal injury was awarded in the past under Nuisance, as in the case of *Hale v Jennings (1938)*, a case involving a 'chair-o-plane' that escaped from the defendant's land and crashed onto the claimant's property. However, recent cases, including *Transco v Stockport Metropolitan B.C. (2004)*, suggest that personal injury is not claimable in Nuisance, including under Rylands v Fletcher. However, compensation for loss of profits may be claimable, and this will be of importance to self-employed persons who lose money because they are incapacitated by an injury caused by the release of toxic substances from the defendant's premises.

The liability under Rylands v Fletcher would originally extend to any losses incurred by the claimant as a result of the escape. However, the more recent case of *Cambridge Water Co. Ltd. v Eastern Counties Leather PLC (1994)* has limited the extent to which the claimant can recover damages under the tort. In this case, the defendants had allowed toxic solvents to pollute the land on the premises from which they had operated. Over time these had drained though the soil and into underground aquifers, from which the claimant company now sought to extract water for public consumption. Due to the pollution by the solvent waste, the water was unfit for this purpose, and the claimant sought to obtain compensation. The claimant failed in his claim because it was held that the defendant could not foresee that his actions would lead to the pollution of underground water supplies, but believed that the solvent would evaporate off the land.

As a result, the liability under Ryland v Fletcher is no longer seen as strict in nature but limited to damages that are 'reasonably foreseeable'. As such, it is no longer clear that there is any difference between the liability here and that in common law Negligence. In addition, the courts have continued to restrict the application of the rule, so that liability for hazardous substances will normally lie in common law Negligence or Nuisance, or will be subject to statutory controls.

Defences

The defences available in the tort of Ryland v Fletcher are very similar to those in Private and Public Nuisance. They include the act of a stranger, default of the claimant, act of God or the operation of statutory authority. As a result, while liability is strict, it is clearly not absolute.

Statutory controls

Introduction

In addition to the common law liability for Nuisance, there are a wide range of statutes that impose liability on the employer and other parties for the creation of nuisances. The key one here is the *Environmental Protection Act 1990*, which replaces the *Public Health Act 1936* and the *Control of Pollution Act 1974*. Most of this legislation relates to environmental matters that are beyond the scope of this book. However, it is important to be aware of the key requirements of the *Environmental Protection Act 1990*.

Environmental Protection Act 1990

The *Environmental Protection Act 1990* deals with various nuisances under Part III of the act. This includes premises in a state prejudicial to health or a nuisance, accumulations or deposits that are prejudicial to health, or accumulations of dust or effluvia that are prejudicial to health. The local authority will take action to deal with these problems in order to protect public health.

The local authority here will enforce the law by way of a notice procedure backed up by prosecution if the defendant fails to take action. In addition, the High Court can issue an injunction if the local authority feels that the statutory system of enforcement is being abused. In many ways, the system is similar to the enforcement of health and safety legislation by the HSE and local authorities, using the enforcement notice system.

Conclusion

While contractual liability is less important in respect of civil liability for health and safety, it should not be ignored. The employment law remedies are important in ensuring that employees can obtain an element of protection against discriminatory action taken by an employer, following action on their part over health and safety matters. The fact that it is possible to take an action in an Employment Tribunal is very important in itself, as such actions are still much less expensive than taking a civil action for Negligence through the normal court system. In addition, the right of the Employment Tribunal to make a declaration as the right of the parties in relation to the matters raised before it may lead to the parties coming to an amicable settlement.

As far as commercial arrangements are involved, the rights and obligations imposed by law under the *SOGA*, *SSGA* and *SGSA* all have the effect of shifting most of the responsibility for goods or services onto the supplier, especially if the end result is an accident causing death or personal injury. Only in the case of non-injury accidents can liability be apportioned between the two parties, and then only if agreed between two parties in a non-consumer contract and if the court believes that the agreement was 'reasonable'. Even then, the liability will usually fall on the supplier, who is usually in a better position to provide insurance cover for such eventualities.

The importance of contractual liability is likely to grow as more employees are engaged in work outside the UK, because the remedy in Negligence is not available. In addition, the contractual agreement may be used increasingly to allocate responsibilities between different parties, so that they can arrange appropriate insurance cover. This will be of importance in the case of contractors and subcontractors, whose relationship is regulated by *SGSA*. However, as noted above, the court will not allow labour-only subcontractors to be treated as anything but employees for purposes of health and safety, so the restrictions of the *Unfair Contract Terms Act 1977* will continue to apply. The key issue here is how to distinguish between genuine self-employment and disguised employment, especially in a rapidly changing labour market.

Liability in Nuisance is more problematic, as it is mainly limited to claims for property damage, rather than personal injury. However, in preventing the ingress of smoke, noise or toxic material onto a person's property, they may also have an impact of health and safety matters if the property is also a place of work. More often, it is the defendant who is operating a place of work, so we are looking at the liability of the employer to third parties. The scope of Public Nuisance and liability under Rylands v Fletcher is certainly of interest, as it allows for a claim for loss of profits, which is not claimable under the Tort of Negligence. Consequently, where claims are made by employees, then the Negligence action will suffice, but where the claimants are engaged as self-employed, then the right to claim for loss of profits could be very useful.

The key issue in the Tort of Nuisance relates to the claims of businesses for damages caused by the release of hazardous substances. The recent case of *Cambridge Water* v *Eastern Counties Leather PLC (1994)* suggests that the strict liability in Ryland v Fletcher no longer applies and that the difference between the Tort of Nuisance and the Tort of Negligence has somewhat narrowed. We will have to wait for future legal developments to see how this situation evolves.

Note

1   See K. Bridges 'All in good order' *Safety and Health Practitioner* 2005

5   Subordinate legislation

Introduction

The *Health and Safety at Work Act 1974 (HSWA)* replaced a wide variety of specialised legislation such as the *Factories Act 1961* and the *Mines and Quarries Act 1954*. However, *HSWA* only lays down very broad legal duties on employers and employees. Even the more specific duties in Section 2(2) of the act only relate to the very broad areas of work equipment and the workplace. The original idea under Lord Robens was that *HSWA* would be the only piece of health and safety legislation, but it would be supported by a wide range of Codes of Practice, which would provide more specific guidance to cover the particular problems in different industries. In the event, joining the EU meant that the UK had to introduce specialised legislation, rather than relying on Codes of Practice, as this was the practice in other EU countries. The regulations are passed under Section 15 of *HSWA* and focus on specific hazards rather than specific industries. As a result, there are regulations to cover such matters as work equipment, workplaces, manual handling, mechanical handling and the control of hazardous substances, among other matters. I refer to this legislation as subordinate legislation, because it is passed under the auspices of *HSWA*, and because it addresses specific issues rather than the general principles of health and safety at work.

In this chapter I would like to look at some of the key pieces of subordinate legislation and how they have been interpreted by the courts. The legislation is not always qualified by the requirement of what is 'reasonably practicable', but may impose strict liability or liability qualified by such terms as 'practicable'. Until recently, most of the regulations allowed for civil liability, so the courts were often required to clarify the meaning of the terms used. However, it is not clear whether the courts will apply a similar approach when the regulations are only used for criminal prosecution, given the unwillingness of the courts to impose strict liability in criminal matters.

In this chapter, I will start by looking at the regulations dealing with safety management, such as the *Management of Health and Safety at Work Regulations 1999 (MHSWR)* and the *Construction (Design and Management)*

Regulations 2007. After that, I will look at the regulations covering general workplace hazards, including the workplace, work equipment and personal protective equipment. In the third part, I will look at hazards related to work activities, such as manual handling, lifting equipment, working at height and the use of display screen equipment. In the final section, I will look at physical hazards and hazardous substances, including electricity, fire, noise, vibration, general hazardous substances, asbestos and lead. It should be noted that many of these regulations do not cover merchant shipping, dockside activities or offshore oil and gas installations, as these sectors have their own specific regulations.

Safety management regulations

There are two main sets of regulations here to consider. There is the *Management of Health and Safety at Work Regulations 1999 (as amended in 2003)*, which covers most of the key safety management requirements. However, there are special requirements in the construction industry, because of the widespread use of subcontracting, as a result of which we have the *Construction (Design and Management) Regulations 2007*, which I will consider separately. In addition, I will cover the *Reporting of Injuries, Disease and Dangerous Occurrences Regulations 2013*, which apply to all workplaces.

The Management of Health and Safety at Work Regulations 1999 (MHSWR)

The *MHSWR* sets out some of the key elements of a safety management system and applies to all work activities covered by *HSWA*. The regulations were passed to implement the EC Directive 89/655/EEC on minimum safety requirements. There is an Approved Code of Practice and Guidance, (L21), entitled 'Management of Health and Safety at Work', but to an extent the regulations should also be read in conjunction with the HSE Guidance note HSG 65, 'Managing for Health and Safety', which is based on the management concept of 'plan, do, check and act'. The *MHSWR* expands on the basic principles of Section 2(1) of *HSWA* by setting down what the employer needs to do to manage his workplace in order to ensure the health and safety of his employees, as well as third parties.

Risk assessments

Regulation 3 sets out the need to carry out a 'suitable and sufficient' assessment of risks to health and safety. This is the starting point for a safety management system, as it enables the employer to prioritise the key safety issues in the workplace. The risk assessment must be reviewed if there is a significant change in the matters to which it relates. Where the employer has five or more employees the findings of such a risk assessment must be recorded. A 'suitable and sufficient' risk assessment means a risk assessment that will identify the key risks arising from the work activity, and which will remain valid for a specified time. Reference may be made here to the HSE Guidance note *Five Steps to Risk Assessment*,[1] which sets out how to carry out a risk assessment.

There is little case law on when a risk assessment is required. However, in *Hall v Edinburgh City Council (1999)* it was suggested that the failure to carry out a risk assessment was itself evidence of a failure to do everything that was 'reasonably practicable'. In contrast, the later case of *Koonjul v Thameslink Healthcare Services NHS Trust (2000)* held that it was not necessary in a small organisation for a risk assessment to be carried out for all activities. The latter approach is to be favoured, otherwise there is a danger that risk assessments will become an excessive burden on industry. A risk assessment should only be required for activities that appear to involve an element of risk.

Safety procedures

Regulation 4 refers to the principles of prevention to be used by the employer, and these are listed in Schedule 1 to the regulations. These principles include such matters as; avoiding risks, evaluating those risks that cannot be avoided, combating risks at source and giving priority to collective over individual protective measures. These general principles are similar to the so-called 'hierarchy of controls' used by many safety organisations. As such, they form the basis of the safety procedures to be implemented.

Regulation 5 requires the employer to implement health and safety arrangements according to the nature of the organisation and its work activities. The ACOP suggests the setting up of a safety management system based on the 'plan–do–check–act' principle. The HSE proprietary safety management system is HSG 65 as referenced above, but it is also possible to use an alternative proprietary safety management system such as OHSAS 18001:2007, as promoted by the British Safety Institute among others.[2]

Regulation 6 sets out the need for employers to provide 'appropriate' health surveillance, in accordance with the risks arising from their work activities. Some work activities will require health surveillance, such as where employees work with lead or other hazardous substances. It is for this reason that the same requirement for health or medical surveillance appears in the relevant regulations appertaining to lead and hazardous chemicals.

Regulations 7 and 8 relate to the development of specific safety procedures in the workplace. Regulation 7 requires the employer to appoint competent persons to help in the implementation of health and safety measures. Such competent persons would include not only a health and safety manager, but also such roles as a first aider, a fire warden or a person in charge of personal protective equipment. Regulation 8 requires the employer to establish procedures to deal with serious and imminent danger, such as the creation of emergency procedures to deal with a fire or the escape of dangerous chemicals.

Information, training and supervision

Regulations 9 and 10 deal with the need to establish effective communications with various parties. Regulation 9 relates to the establishment of contacts with emergency services, continuing on from the development of emergency procedures in Regulation 8. Regulation 10 covers the need to provide employees with 'comprehensible and relevant information' on the risks to their health and safety. This develops the requirement in Section 2(2)(c) of *HSWA* and refers to the need to provide information on risks to health and safety identified by a risk assessment, as well as the preventative measures to be implemented in order to provide protection against such risks.

Regulation 11 requires employers sharing a workplace to co-operate with each other in co-ordinating any safety measures, as well as informing employees of the other employer of any risks affecting their health and safety. In addition, Regulation 12 requires an employer to provide comprehensible information on risks to health and safety to the employees of any outside organisation working on their undertaking. This will help to ensure the safety of subcontractors working on their premises, and in many ways is a consequence of the outcome of the earlier case of *R* v *Swan Hunters Ltd (1982)* brought under Sections 2(2)(a), 2(2)(c) and 3(1) of *HSWA*. Finally, Regulation 13 deals with the duty on the employer to ensure that his employees are provided with adequate health and safety training, both on recruitment and in the case of exposure to any new risks.

Duties of employees

Regulation 14 imposes duties on the employee, which arguably go well beyond those introduced in *HSWA*. Under Section 7 of *HSWA* the employee has a duty to take reasonable care of himself and other persons and to co-operate with the employer in helping him perform any duties under the act. However, Regulation 14 of *MHSWR* requires the employee to be more pro-active. He is required to use any work equipment and safety devices in accordance with his training and instruction. Moreover, it is now his duty to bring to the attention of his employer any work situation that he would reasonably consider to be a serious and imminent danger to health and safety, or a shortcoming in his employer's protection arrangements.

Protected employees

Regulations 15 to 19 deal with specific additional protections for temporary workers, young persons and expectant mothers. Under Regulation 15, temporary workers have to be provided with comprehensible information on the skills required to carry out the work, any health surveillance required and any special features of the job affecting their health and safety. Under Regulation 16, new or expectant mothers need to have a specific risk assessment and they may be restricted from carrying out certain jobs. Similar restrictions are placed on the employment of young persons under Regulation 19.

Defences to liability

Regulation 21 states that in criminal proceedings it is no defence for an employer to argue that the contravention was due to a fault on the part of an employee or person appointed by him under Regulation 7. This underpins the principle of vicarious liability for criminal matters and was introduced to overrule the decision in *R v Nelson Group Services (Maintenance) Ltd (1999)*. However, it does not prevent an employer arguing that he had done everything 'reasonably practicable' to avoid commission of an offence so that liability under *HSWA* would be avoided, as in the later case of *R v Hatton Traffic Management (2006)*.

Civil action against employees

Following the amendments to the regulations in 2003, it was possible for a third party to take a civil action against an employer or an employee under the *MHSWR*. However, this posed a problem for employees in respect of any duty under Regulation 14, as it was felt unfair to place a liability on an uninsured party. As a result, liability in respect of employees was abolished by virtue of the *Management of Health and Safety at Work (Amendment) Act 2006*.

Summary

We can see that the Regulations are very wide in scope and it is likely that they will be used by the enforcement authorities as an addition to a breach of *HSWA*, especially in the case of a clear failure of a safety management system.

Construction (Design and Management) Regulations 2007

The Construction (Design and Management) Regulations 2007 (CDM) implemented the requirements of the EU Directive 92/57/EEC. The regulations are split into two main areas, Parts 2 and 3 set out the management duties on a construction site and are an extension of the *MHSWR*. The other area is Part 4, which sets out key health and safety requirements on construction sites. There is an Approved Code of Practice and Guidance Notes, (L144), entitled 'Construction (Design and Management) Regulations'.

Site management issues

The regulations were passed to improve the management controls on construction sites, which are characterised by the widespread use of sub-contracting. The regulations set out the main management responsibilities of five key parties: the client, the designer, the CDM co-ordinator, the principal contractor and the subcontractor. The nature of these parties is defined in Regulation 2, which also sets out the requirements of two key documents to be produced by the parties, namely the health and safety plan and the health and safety file.

Regulation 4 requires the person appointing the key parties to ensure that they are competent, while Regulation 5 requires the parties to co-operate with each other, to ensure that they can perform their allotted duties and functions under the regulations. If the project is notifiable to the HSE, there are additional duties such as the appointment of the CDM Co-ordinator. The main liabilities of the key parties are as set out below.

THE CLIENT

The client is the party who seeks to carry out the building project or accepts the services of another to achieve the same, and as such a client may be a business or an individual. Under Regulation 9, the client is responsible for ensuring that the project is designed in a safe manner and is carried out safely. The client will appoint the designer, CDM Co-ordinator and the principal contractor, although he may obtain the services of an agent to help him do this if he does not have sufficient expertise, or he can even rely on the CDM Co-ordinator. The client is also responsible for providing information about the site to the above parties and for ensuring that there are sufficient resources to enable the construction project to be carried out safely. He will provide a copy of the health and safety file for the building if there is not one already drawn up.

THE DESIGNER

The designer is the party who designs the building and will usually be an architect. Under Regulation 11 he is responsible for ensuring that the design of the building is safe, both in relation to its construction and the carrying out of maintenance activities. To this extent, he should take into account the duties laid down in the *Workplace (Health Safety and Welfare) Regulations 1992.*

THE CDM COORDINATOR

The CDM Co-ordinator is the person appointed to co-ordinate the construction activities of the various parties on behalf of the client, under Regulation 14(1). The duties of the CDM Co-ordinator are various and include notifying the HSE that the project is regulated by CDM. He has to advise the client on how to comply with his duties under the *CDM Regulations* and he needs to liaise with other parties to ensure that suitable arrangements are made to co-ordinate health and safety measures during the planning stage of the project. He will prepare a health and safety file for the new construction, if one does not already exist.

THE PRINCIPAL CONTRACTOR

The principal contractor is appointed by the client under Regulation 14(2) and his duties are laid down in Regulations 22 to 24. These duties include the management of the construction phase of the project, to ensure that health and safety is maintained. This includes liaising with the key parties above and exercising control over any subcontractors appointed to help him carry out the project. It is his duty to prepare the construction phase plan and to review and revise it as appropriate.

SUBCONTRACTORS

The general duties of a subcontractor are set out in Regulation 13 and include the need to plan, manage and monitor any construction work that he carries out, or which is carried out under his control, so as to ensure, so far as is reasonably practicable, that the work is carried out without risk to health and safety.

SAFETY PLAN AND SAFETY FILE

The safety plan sets out the procedures to ensure that the project can be carried out without risk to health and safety. The pre-construction plan is provided by the client and contains information on the nature of the site, the proposed use of the structure and the time scales for planning and preparation work. The construction phase plan is drawn up by the principal contractor and covers how to carry out the live project, ensuring that the work is effectively planned, managed and monitored. The health and safety file contains information that might be needed when carrying out any future construction work or modifications. The file is commenced by the client but updated by the CDM Co-ordinator.

The above regulations set out a comprehensive management structure to ensure the health and safety of employees in a construction project. It can be seen that the need for effective planning and documentation is a key issue in the development of such a safety management system.

General safety requirements

In addition to the general management controls, there are a number of general safety requirements, as set out in Part 4 of the regulations.

Regulation 26 provides for suitable and sufficient safe access to, and egress from, every place of work. This will cover access to heights and the duty is qualified by the term 'reasonably practicable'. Other regulations are more specific, referring to the need to ensure that the site is kept in good order and secure (Regulation 27), while Regulation 29 refers to the safe planning of demolition activities, and Regulation 30 to the safe storage, transport and use of explosives.

Regulation 31 deals with measures to ensure safety in the carrying out of excavations, in order to prevent a collapse, while Regulation 32 controls the construction of cofferdams and caissons.

After these specific issues, the rest of the regulations refer to general problems that may occur on any site, construction or otherwise. They include the control of traffic routes (Regulation 36) and control over the movement of vehicles (Regulation 37). There is the provision of emergency procedures (Regulation 39) and measures to detect and deal with fires on the site. Other regulations deal with protection from extreme weather conditions (Regulation 43) and the provision of adequate lighting (Regulation 44).

Reporting of Injuries, Diseases and Dangerous Occurrences Regulations 2013

These regulations set out the rules for the reporting of specific accidents, diseases and dangerous occurrences to the enforcement authorities. Regulation 2 defines an accident as including 'an act of non consensual physical violence done to a person at work'. A disease is defined as including a 'medical condition', while a 'dangerous occurrence' is one of a number of occurrences arising out of work as set out in Schedule 2 to the regulations. Regulation 3 defines who is deemed to be a responsible person for the purposes of reporting an injury, death or dangerous occurrence.

Under Regulation 4(1) the enforcement authority has to be informed by the quickest practicable means in the case of a number of non-fatal injuries, including a fracture to any bones (excluding the finger, thumb or toe), amputation, blinding, or crush injuries to the head or torso, as well as any injury that causes loss of consciousness or requires the person to be admitted to hospital for more than 24 hours. This must be followed up by a written notification within ten days. In addition, under Regulation 4(2) the enforcement authorities need to be informed in writing within fifteen days of any accident, which means that an employee is unable to return to work or to his normal work activities for a period of seven days or more.

Under Regulation 5, where a person who is not at work suffers a work-related accident and is admitted to hospital, then that has to be reported in the same way as under Regulation 4(1). In addition, Regulation 6 requires the same procedure to be followed in the case of a fatal work-related accident, which includes death as the result of occupational exposure to a biological agent. Regulation 7 refers to dangerous occurrences, which must also follow the same reporting procedure as above.

Regulation 8 lists a number of occupational diseases that need to be reported to the enforcement authorities. These include carpal tunnel syndrome, occupational dermatitis, hand arm vibration syndrome, occupational asthma and tendonitis. These must be reported under the reporting procedure laid down in Regulation 4(1), in other words reporting by the quickest practicable means followed by written notification within ten days.

There are three schedules to the Regulations, which give further details. Schedule 1 sets out the reporting and recording procedures, while Schedule 2 sets out the types of dangerous occurrences reportable under Regulation 7 and Schedule 3 lists the reportable diseases referred to in Regulation 8.

General Safety Regulations

Under this category I want to look at three sets of legislation, which are general in nature; namely the *Workplace (Health, Safety and Welfare) Regulations 1992 (WHSWR)*, the *Provision and Use of Work Equipment Regulations 1998 (PUWER)* and the *Personal Protective Equipment at Work Regulations 1992 (PPE Regulations)*. Effectively, these regulations took over the bulk of the requirements of the old Factories Act 1961. While *WHSWR* concentrates on the workplace itself, including access and egress, *PUWER* and the *PPE* Regulations focus on the equipment used within. All of these regulations have been subject to considerable judicial interpretation, mainly in the light of civil actions taken under them. Although there is now only criminal liability, the courts will still use the same cases to interpret the wording of the regulations, even when used in criminal prosecutions.

The Workplace (Health, Safety and Welfare) Regulations 1992

Scope of the regulations

The *WHSW Regulations* were passed to implement the EU Directive 89/654/EEC. They were amended in 2002 and are supported by an Approved Code of Practice (ACOP). The regulations now apply to almost all workplaces, with the exception of quarries and construction sites, which are covered by their own legislation. The duties cover all workers, but they do not extend to non-workers, as made clear in the civil case of *Donaldson* v *Hays Distribution Service Ltd (2005)*. The liability in the legislation is sometimes strict, but it may also be qualified by the term 'reasonably practicable'.

Structure of the workplace

Regulation 4A requires that any workplace building should have a stability and solidity, appropriate to the nature of the activity being carried on there. Regulation 5 goes on to state that the premises and any equipment on the premises must be maintained to ensure they are in 'an efficient state, in efficient order and in good repair'. These regulations impose a strict liability and have been seen as effectively limited to structural matters rather than transient defects, which come under Regulation 12.[3] The equipment at issue here will be equipment that is integral with the building, such as ventilation systems and emergency lighting, as opposed to work equipment and machinery, which would be covered under *PUWER*.

Workplace environment

Regulation 6 provides for ventilation that is 'effective and suitable', again a strict liability. The HSE Guidance Note (HSG 202), entitled 'General Ventilation in the Workplace: Guidance for Employers', states that there should be at least five to eight litres of fresh air per second per occupant. Regulation 7 requires the provision of a reasonable temperature in the workroom, so as to enable persons to work in reasonable comfort. The ACOP states that the temperature should be 16 degrees Centigrade for normal working, and 13 degrees Centigrade where the work involves considerable physical effort. Although no maximum temperature is specified, it is arguable that a temperature above 25 degrees Centigrade would be evidence of inadequate ventilation.

Regulation 8 sets out the need for 'suitable and sufficient lighting' and this may involve local lighting where necessary. What is suitable and sufficient will depend on the level of light needed to ensure that the work is carried out safely, and various levels are suggested for different activities.[4] Regulation 9 covers the need to keep the workplace clean, while Regulation 10 sets out minimum space requirements. These are specified in more detail in Schedule 1 Part I to the regulations and require any existing workplace to provide for at least 11 cubic metres per person.

Workstations and traffic routes

Regulation 11 sets out the requirements for workstations. Where they are located outside, they need to provide protection from the elements and allow for speedy egress in case of emergency. Wherever it is possible to perform the job sitting down, suitable seating must be provided. The regulations are supplemented by the *Health and Safety (Display Screen Equipment) Regulations 1992*, which provide for specific requirements where a person habitually uses display screen equipment as part of his job.

Regulation 12(1) requires the surface of all floors and traffic routes to be constructed so as to be suitable for use. This requirement would appear to be strict and such surfaces should not be overloaded, as per *Grieves* v *Baynham Meikle (1975)*. Regulation 12(3) requires every floor and traffic route to be kept clear of any obstruction or any article or substance that may cause a person to slip, trip or fall. This requirement is qualified by the term 'reasonably practicable'. However, the regular presence of water on the floor may be seen as constituting an unsuitable floor surface and so fall under Regulation 12(1), as in *Ellis* v *Bristol City Council (2007)*, where an employee was injured by slipping in a pool of urine on the floor of a care home. However, the court will not find an employer liable for a sudden and unexpected event, as made clear in *Lewis* v *Avidan (2005)*.

Regulation 17 looks at the subject of traffic routes, requiring them to be organised in such a way as to ensure that pedestrians and vehicles can circulate safely. This should ideally involve segregation, but if not, it still requires suitable measures to ensure safety. There is relevant guidance in the HSE publication (HSG 136), entitled 'Workplace Transport Safety: Guidance for Employers'.

Welfare facilities

The final regulations relate to welfare matters. Regulation 20 covers the requirement for sanitary conveniences and Regulation 21 for washing facilities. In both cases the provision is for 'suitable and sufficient' facilities. In the case of sanitary conveniences, Schedule 1 Part II to the act stipulates one suitable water closet for every 25 persons of either sex. Regulation 22 refers to the need to provide 'wholesome drinking water' that is readily accessible to staff. Regulation 23 requires the provision of accommodation for clothing, while Regulation 24 provides for 'suitable and sufficient' changing facilities where staff have to change into work clothes. Finally, Regulation 25 provides for rest facilities and a place to eat meals, separate from the workplace.

Summary

Overall, the regulations provide a comprehensive list of requirements including building safety, environmental and welfare issues. The liability is sometimes strict, but in the case of existing buildings it is often modified to a 'reasonably practicable' requirement. However, as the wording in the regulations is often a little vague, supporting guidance notes are often essential in clarifying the specific nature of the requirements.

Provision and Use of Work Equipment Regulations 1998 (PUWER)

Scope of the regulations

The *Provision and Use of Work Equipment Regulations 1998* implement EU Directive 89/655/EC. The regulations replace part of the *Agriculture*

(Safety, Health and Welfare Provisions) Act 1956, the *Factories Act 1961*, *the Offices Shops and Railway Premises Act 1963*, the *Power Presses Regulations 1965*, and the *Woodworking Machines Regulations 1974*.The regulations have been subject to considerable statutory interpretation in decided cases and although almost all of these are civil cases the same interpretation should apply in criminal cases. The regulations are accompanied by an Approved Code of Practice and Guidance Note (L22), entitled 'Safe Use of Work Equipment'.

Regulation 2 is an interpretation section and this defines the term 'work equipment' as being 'any machinery, appliance, apparatus or tool or installation for use at work'. This definition has been widely interpreted to include such diverse objects as a ship in *Coltman* v *Bibby Tankers (1988)* or even a paving slab, in *Knowles* v *Liverpool City Council (1993)*. In the case of *Hammond* v *Commissioner of Police for the Metropolis (2004)* the court took a very restricted definition of work equipment, differentiating it from the work piece. However, the later case of *Spencer-Franks* v *Kellogg, Brown Root Ltd (2008)* widened this to include any object that performs a useful or practical function at work, in this case a self-closing door on an oil rig.

Regulation 3(2) states that the regulations apply to an employer, in respect of equipment provided for use or used by an employee. Under Regulation 3(3)(a) this is widened to include self-employed persons, and under 3(3)(b) to include any person with control over work equipment or the person who uses it. The scope of the regulation covers the carrying on of a trade or business, whether or not for profit, but does not include domestic use.

Provision of work equipment

Regulation 4 requires every employer to ensure that work equipment is 'so constructed or adapted as to be suitable for the purpose for which it is used or provided'. The term 'suitable' is itself further defined in regulation 4(4) as 'suitable in any respect which it is reasonably foreseeable will affect the health and safety of any persons'.

As such the liability is not strict, but involves an element of fault-based liability similar to common law Negligence. What exactly is 'reasonably foreseeable' will be determined by the courts and this has led to some interesting decisions. In *Wallace* v *Glasgow City Council (2011)* it was decided that it was 'reasonably foreseeable' that an employee would stand on a toilet bowl to open a toilet cubicle window and so would be able to claim when the toilet bowl broke and claimant slipped off. Similarly, in *Robb* v *Salamis (2006)* the court held that a portable ladder for a bunk bed was not suitable for its purpose, because it might not be properly replaced. As such, it was held 'reasonably foreseeable' that the employer should anticipate such a situation, which might give rise to an accident occurring.

Maintenance and inspection

Regulation 5 requires every employer to ensure that work equipment is maintained in 'an efficient state, in efficient working order and in good repair'. It was this regulation that was used in the seminal case of *Stark* v *Post Office (2000)*. The employer was held liable for the sudden failure of a calliper on the claimant's delivery bicycle, which led to an injury. The fact that the employer regularly checked the condition of the bicycle was not sufficient to negate liability, making it clear that liability here was effectively absolute. Note here the later case of *Smith* v *Northamptonshire County Council (2009)*, where it was held that there was no liability in respect of equipment that was not in the control of the employer and had not been supplied by him, in this case a wooden ramp providing access to the home of an elderly client.

Regulation 6 requires the employer to inspect work equipment where its safety depends on the 'installation conditions'. This includes inspecting the equipment after installation, when assembled at a new site and where subject to deterioration, carrying out inspections at regular intervals.

Information and training

Regulation 8 refers to the need to provide information and instruction to persons using work equipment, while Regulation 9 covers training for such persons. The need for adequate training was emphasised in the case of *Allison* v *London Underground Ltd (2008)*, where the claimant developed tenosynovitis as a result of using a new traction brake controller on an underground train. The court held that the duty of the employer to provide adequate training under the regulations was a stricter duty than in common law Negligence, which was restricted to those risks that were 'reasonably foreseeable' to the defendant. Instead, under Regulation 9, the duty required that the employer should consult suitably qualified persons to ensure that the training was adequate in the circumstances.

Machinery safety

Regulation 10 requires work equipment to conform to EU requirements in respect of design and construction. The main directive here is the Machinery Directive, now implemented in the UK by way of the *Supply of Machinery (Safety) Regulations 2008*.

Regulation 11 of *PUWER* requires employers to ensure that measures are taken to prevent access to dangerous parts of machinery, or to stop such machinery before a person enters the danger area. This regulation is similar to Section 14 of the *Factories Act 1961* and introduces the key guarding arrangements. The above regulations refer to the need for either a fixed or interlocked guard or some other safety device, such as a trip switch to cut off power when the employee enters the danger zone.

Any such guard or protective device has to be suitable for the purpose for which it is provided, of good construction, maintained in an efficient state and efficient working order, and should not give rise to any increased risk to health and safety. The guarding requirements are supplemented by the relevant ACOP, which shows how to carry out guarding for various types of machinery.

Regulation 11 only deals with preventing access to the dangerous machinery, but Regulation 12 goes further, and requires the employer to take measures to prevent risks to health and safety arising from exposure to other hazards involving the use of work equipment. The type of hazards here include the ejection of articles or substances from work equipment, the disintegration of work equipment, the discharge of any article or gas and the possibility of explosion. This Regulation dealt with the gap in the previous *Factories Act 1961*, as revealed by the case of *Close v The Steel Company of Wales Ltd (1962)*. Here the employee was injured when a drill exploded, but he was unable to claim compensation under Section 14 of the *Factories Act 1961* because it did not cover the ejection of material injuring the worker.

Controls and isolation procedure

Much of the rest of the Regulations relate to the various controls required to ensure employee protection. So Regulation 14 covers the need for starter controls and controls to alter the speed of the process, while Regulation 15 covers stop controls. Regulation 17 requires any controls to be clearly visible and identifiable, to ensure that they can be reached by the operator of the machine when required. As such, Regulation 17 is really dealing with ergonomics as regards the design of machinery controls and control panels. Finally, Regulation 18 requires all control systems to be safe, as far as is reasonably practicable.

Regulations 19 to 24 then deal with associated issues, such as the ability to isolate the equipment (Regulation 19), the stability of the equipment (Regulation 20), the need for 'suitable and sufficient lighting' (Regulation 21), the provision of adequate markings (Regulation 23) as well as warning signs or devices (Regulation 24).

Summary

The above regulations cover the whole area of machinery guarding and controls for a wide range of equipment. In addition, there are specific sections dealing with the requirements of mobile work equipment including fork lift trucks (Regulations 25–30), as well as power presses (Regulations 31–35).

Personal Protective Equipment at Work Regulations 1992

The *Personal Protective Equipment at Work Regulations 1992* implemented the EU Directive 89/656/EEC entitled 'Minimum Health and Safety Requirements for the Use of Personal Protective Equipment'. Civil liability here had long imposed a duty on the employer to provide Personal Protective Equipment (PPE) and to ensure it was worn, as in the case of *Bux v Slough Metals Ltd (1974)*. The statutory Regulations are accompanied by a Guidance Note (L25), entitled 'Personal Protective Equipment at Work', which provides useful information on how to comply with the regulations.

Scope of the regulations

Regulation 2 defines PPE as 'all equipment (including clothing affording protection against the weather) which is intended to be worn or held by a person at work, and which protects him against one or more risks to his health or safety'. It should be noted here that whereas the PPE may be specific to one of the risks encountered, it need not be suitable for other risks. So in *Fytche v Wincanton Logistics PLC (2008)* the claimant suffered frostbite due to freezing water leaking through a hole in the safety boots provided by his employer. Nevertheless, the employer was held not liable, as the boots were designed to protect his feet from being injured by falling loads and not to protect them from inclement weather.

Regulation 3 specifies that the regulations do not cover ordinary working clothes, which do not offer any protection for the wearer from health and safety risks. Moreover, the regulations do not apply where alternative specialist regulatory provision exists, such as the *Control of Lead at Work Regulations 2002*, the *Control of Substances Hazardous to Health Regulations 2002*, the *Control of Noise at Work Regulations 2005*, and the *Control of Asbestos Regulations 2012*.

Duties of the employer

Under Regulation 4 the employer is required to ensure that 'suitable' PPE is provided to employees, who may be exposed to a risk to their health and safety at work. In determining what is 'suitable', Regulation 4(3) states that it should be 'appropriate' for the risks involved, the condition of the workplace and the period for which it is expected to be worn. It should also take into account the ergonomic requirements of the wearer, including being capable of fitting correctly. In effect, it should 'so far as is practicable' be effective to prevent or adequately control the risks involved, without creating a greater overall risk. What is appropriate PPE was considered in the case of *Threlfall v Hull City Council (2010)*. Here it was held that gloves provided by the employer were not adequate because they failed to protect the employee from being injured by a sharp object when collecting bags of rubbish. In effect, the liability was determined by the occurrence of the accident, as such imposing a form of strict liability.

The duty here is similar to that created under *PUWER* in respect of work equipment. It should be noted that the provision of PPE is still seen as a last option, as made clear in Regulation 4(1), which states that the provision of PPE is only required to the extent to which the risk has not been controlled by other means, which are equally or more effective.

Regulation 5 requires the employer to ensure that where more than one type of PPE is worn, the PPE provided is compatible and continues to provide effective protection. A problem here arises from the wearing of safety glasses with hearing protectors, as the glasses will often undermine the effectiveness of the hearing protection by breaking the seal with the face. In some cases, integrated hearing and eye protection can be provided.

Risk assessment

Regulation 6 requires the employer to assess the risks to health and safety of the employee and to ensure that the correct PPE is provided. In this respect, it is important to take into consideration the personal characteristics and disabilities of the employee, as this may affect the level of risk. An early example here in common law is the case of *Paris v Stepney Borough Council (1951)*. In this case, the employer was liable in Negligence when he failed to ensure that an employee with one good eye wore safety goggles, even though an employee with two good eyes would not normally be expected to wear such protection when carrying out that particular job.

Maintenance of PPE

Under Regulation 7 there is an obligation on the employer to maintain PPE to ensure that it is 'in an efficient state, in efficient working order and in good repair'. This obligation would appear to impose a strict liability, as it is couched in identical language to that used in Regulation 5 of *PUWER*, which provides the basis for the decision in *Stark v Post Office (2000)*.

Information, instruction, training and supervision

Regulation 9 requires the employer to provide information, instruction and training, so as to enable the employee to be aware of the risks that the PPE is designed to deal with, as well as the correct way in which to use the PPE and to ensure that it is properly maintained. In addition, Regulation 10 requires the employer to supervise the employee, so as to ensure that he uses the PPE provided in accordance with that training.

Summary

The regulations are in many ways similar to *PUWER*, and that is not surprising given that Personal Protective Equipment is still effectively a form of work equipment. The main difference is the need to carry out an effective risk assessment to determine the correct PPE to be used, both in respect of the risk to be dealt with and the person to be protected.

Activity based safety regulations

In this section I want to look at those regulations that relate to a specific work activity. In this respect, I will be looking at four regulations; namely the *Manual Handling Operations Regulations 1992 (MHOR)*, the *Lifting Operations and Lifting Equipment Regulations 1998 (LOLER)*, the *Work at Height Regulations 2005 (WAHR)*, and the *Health and Safety (Display Screen Equipment) Regulations 1992*. These regulations cover a variety of hazards in very different types of work activity and, as such, they have generated some useful case law.

Manual Handling Operations Regulations 1992 (MHOR)

The *MHOR* were passed to implement EU Directive 90/269/EEC and they are accompanied by an Approved Code of Practice, as well as a Guidance Note (L23), entitled 'Manual Handling'.

Scope of the regulations

Under Regulation 2 manual handling operations are defined very widely as 'any transporting or supporting of a load (including the lifting, putting down, pushing, pulling, carrying or moving thereof) by hand or by bodily force'. As such, kicking an item or supporting it on your head would also appear to be covered, and any injury arising would be evidence of a breach of the regulations. However, it is made clear that this does not include an injury caused by toxic or corrosive substances that have leaked from the load, as this would more properly fall within the scope of the *COSHH Regulations*.

Duties of the employer

In Regulation 4 the duties of the employer are set out, these being qualified by the term 'reasonably practicable'. They include, first, seeking to avoid the need for employees to undertake manual handling. If this is not possible, then the employer has to carry out a 'suitable and sufficient' assessment of the operation. On the basis of this, he then needs to take steps to reduce the risk of injury, in respect of such operations, to the lowest level reasonably practicable. In determining the level of risk involved, the employer needs to take into consideration such matters as the physical suitability of the employee, the clothing and footwear he is wearing, the level of experience and training, as well as the results of any risk assessments already carried out.

It is not always possible to eliminate all of the risks involved in manual handling, as made clear in the case of *King* v *Sussex Ambulance NHS Trust (2002)*. In this case, an employee was injured when carrying a patient downstairs in a carry chair when one of the other persons carrying it lost his grip, causing all the weight to fall on the claimant. However, if the employer does not take enough steps to reduce the risk to the lowest level reasonably practicable, then he will be liable, as in the case of *Gravatomb Engineering Systems Ltd* v *Raymond Parr (2007)*.

Duty of the employee

The duty of the employee is set out in Regulation 5 and this involves making full and proper use of any system of work provided for his use by the employer, in compliance with Regulation 4.

Lifting Operations and Lifting Equipment Regulations 1998 (LOLER)

Scope of the regulations

The *LOLER* were passed to implement EU Directive 89/665/EEC and deal with the whole area of lifting operations, including the lifting of people and goods. The regulations are accompanied by an Approved Code of Practice and a Guidance Note (L113), entitled 'Safe Use of Lifting Equipment'. The regulations usually impose strict liability, although in the case of Regulations 5 and 6 they are qualified by the term 'so far as is reasonably practicable'.

Regulation 2 defines lifting equipment as 'work equipment for lifting or lowering loads (including) attachments for anchoring, fixing or supporting it'. As such, lifting equipment will include a wide variety of equipment such as cranes, hoists, vehicle lifting tables and vehicle tail lifts. The term 'lifting operations' is defined in Regulation 8(2) as 'an operation concerned with the lifting or lowering of a load'.

Under Regulation 3, liability is imposed on the employer in respect of lifting equipment used at work, both by a self employed person, or a person who has control over lifting equipment, or control over a person who uses or supervises the use of such equipment.

Condition of lifting equipment

Under Regulation 4, there is a duty to ensure that the lifting equipment is of adequate strength and stability for each load, and this includes the lifting accessories such as shackles, chains, ropes or lifting slings. Liability here is strict, and it is necessary to take into account the interaction of the various forces that the lifting equipment may be subject to.

Lifting of persons

Regulation 5 refers to lifting equipment used to lift persons. Such equipment should be constructed to ensure, so far as is reasonable practicable, that it does not cause a person to be crushed, trapped or stuck, or fall from the carrier. This will include the need to provide suitable devices to prevent the fall of the carrier, such as an extra strong rope or chain, and such a rope and chain should be inspected on a daily basis. Clearly the danger of a fatal accident is the main concern here, hence the need to provide extra protection and to carry out regular inspections. In 2006, for example, BUPA Care Home Ltd were prosecuted over the death of an elderly resident who slipped from a hoist being used to get her out of a bath.[5] It should be noted that the regulation does not cover lifting devices normally used for lifting goods, which are used to lift persons in exceptional circumstances, such as fork lift trucks.

Lifting operations

Regulation 6 requires all lifting equipment to be so positioned or installed as to reduce, as much as is reasonable practicable, the risk of the equipment or load striking a person, drifting, falling or being released unintentionally. This involves locating equipment in such a position as to avoid the need to lift over the top of people. Continuing on from this, Regulation 8 requires all lifting operations to be properly planned and supervised so that they are carried out in a safe manner. A recent case involving a breach of this regulation is *Berry v Star Autos Ltd and Others (2011)*.[6] Here the claimant suffered severe electrical burns, due to his crane or its load coming into contact with live electric cables while lifting an accommodation unit into position.

Maintenance and inspection

Regulation 7 states that lifting equipment must be marked with its safe working load, as well as whether it is designed to be used to lift persons. Regulation 9 sets up a regime of thorough examination and testing of lifting equipment, including lifting accessories. Lifting equipment has to be thoroughly examined before it is put into use for the first time, after installation and after assembly at a new site. In addition, it must be thoroughly examined when subject to any conditions that may cause it to deteriorate, so as to become dangerous. Such an examination should be carried out at least every six months, if it is lifting equipment used for lifting persons or if it is a lifting accessory. In other cases, the examination should be carried out at least every twelve months, or where exceptional circumstances occur that are liable to affect the safety of the equipment.

Under Regulation 10 any defects raised by the examination should be reported to the employer, or any self employed person responsible for the equipment. In addition, the relevant enforcing authority has to be informed if the defect is seen as one that, in the opinion of the examiner, could involve an imminent risk of serious personal injury.

Summary

These regulations create a comprehensive system of controls, which impose strict liability in respect of the operation and inspection of lifting equipment. This reflects the fact that such equipment is seen as very dangerous, as it has the potential to cause fatalities or very serious injuries.

Work at Height Regulations 2005 (WAHR)

The *WAHR* implements a number of EU Directives (2001/4/EC, 89/391/EEC and 92/57/EEC) and replaces a number of sections of the *Factories Act 1961* and regulations of the *WHSWR*. Working at height has long been recognised as one of the most dangerous work activities, responsible for the majority of fatalities at work.

Scope of the regulations

Regulation 2 defines working at height quite widely, to include 'work in any place, including a place at or below ground level'. It also includes obtaining access to or egress from such a place of work, except by way of a staircase in a permanent workplace, if this could lead to a fall from height liable to cause personal injury.

Duties of the employer

Regulation 3 places a duty on an employer in respect of a breach by himself, or any of his employees or persons under his control. Similarly a self-employed person will be liable in relation to work carried out by himself or persons under his control.

Planning of work at height

Under Regulation 4 the employer has to ensure that work at height is properly planned, appropriately supervised and carried out in a manner that is safe, so far as is reasonably practicable. In ensuring this, he needs to take into consideration the selection of appropriate work equipment, planning for emergencies and the nature of the expected weather conditions. Failure to plan accordingly will be a breach of the Regulations, as was found in the case of *Proctor* v *City Facilities Management Ltd (2012)*, where the claimant was injured falling from a ladder that was unsecured.

Control measures

Regulation 5 requires the employer to ensure that any person engaged in the activity of working at height, including the planning and supervision thereof, has the appropriate level of competence or is supervised by such a person. Regulation 6 lays down the means to avoid risks from working at height. This involves taking account of the findings of an appropriate risk assessment, following on from which the employer should ensure that work is not carried out at height, where it is reasonably practicable for it to be done safely at ground level. Otherwise, the employer has to take suitable and sufficient measures to prevent falls of persons, or provide sufficient work equipment (such as safety harnesses or airbags) to minimise the distance or consequence of any such fall. It is clear that the employer is in breach of the regulations if he allows work to continue at height when it is not necessary. This can be seen in the recent case of *Pate's Curator* v *Stewart Homes (Scotland) Ltd (2013)*, where the claimant fell from height while carrying out a job that should have been done at ground level.

Regulation 7 requires the employer to prioritise collective protection measures when selecting work equipment for use at height. The equipment chosen should take into consideration the working conditions, the risks to safety of persons at the workplace, the distance of any potential fall, the duration and frequency of use of the equipment and any additional risks posed by its use. Regulation 8 refers to the requirement for more specific equipment such as guard rails, toe boards, working platforms, nets, airbags and personal fall protection systems, which are further set out in various schedules to the regulations. An inspection regime for most of these items is referred to in Regulation 12 and includes weekly inspection of any working platform from which a person could fall two metres or more. Similarly, any work equipment exposed to conditions which could lead to deterioration that might compromise safety are required to be inspected at suitable intervals, as well as whenever exceptional circumstances occur that may jeopardise the safety of the equipment.

Specific hazards

Regulation 9 requires the employer to take steps to prevent persons falling through fragile surfaces, including the provision of platforms, coverings, guard rails or other similar means of support. In addition, Regulation 10 requires the employer to take suitable and sufficient steps to protect persons against injury from falling material or objects, so far as is reasonably practicable. Finally, Regulation 11 requires the employer to prevent unauthorised access to areas where there is a danger of falling or being struck by falling objects.

Duties on the employee

Regulation 14 imposes duties on persons working at height, which are very similar to those imposed on employees in Regulation 14 of *MHSWR*. They require that such persons report any activity or defect relating to working at height, which is likely to endanger life. Similarly, that person is required to use any work equipment or safety device provided for work at height in accordance with any training provided.

Summary

The *WAHR* provide a comprehensive set of controls to deal with this high risk activity. The key problem is really the ability of the employer to recognise the relevant dangers in advance and this is why it is so important to plan and organise work at height, as set out in Regulation 4.

Health and Safety (Display Screen Equipment) Regulations 1992

Scope of the regulations

The *Health and Safety (Display Screen Equipment) Regulations 1992 (DSE Regulations)* implement the EU Directive 90/270/EEC and are supplemented by the Guidance Note (L26), entitled 'Display Screen Equipment at Work'. The regulations deal with the health problems arising from the rapid growth in the use of display screen equipment in the 1980s.These include the increase in repetitive strain injuries, such as carpal tunnel syndrome, as well as the development of eyesight problems and work related stress.

Regulation 1 defines 'display screen equipment' quite widely as 'any alphanumeric or graphic display screen, regardless of the display process involved'. As such, the definition will include check-out displays and control panels, although it does not include control cabs for vehicles or machinery, or display screens on board a means of transport. The term 'workstation' is defined as an assembly comprising the display screen equipment, any accessories, the desk, chair, printer and other peripheral equipment. A user is an employee (including agency workers) who habitually uses display screen equipment as a significant part of his normal work, this being further clarified in the guidance. An operator is a self-employed person who habitually uses such work equipment, and he is provided with more limited protection.

Workstation assessment

Regulation 2 requires the employer to carry out a 'suitable and sufficient' analysis of the workstation, effectively a risk assessment. As a result, the employer is required to reduce the risks identified to the lowest extent reasonably practicable, for the benefit of both users and operators. More specifically, under Regulation 3, the employer must ensure that the workstation meets the requirements laid down in the schedule to the regulations. This schedule covers a wide variety of matters, such as the design of the screen, keyboard, chair and desk, as well as providing guidance on environmental requirements such as space, light, heating and the control of noise.

Control measures

Under Regulation 4 the employer has to ensure that users are given periodic breaks or changes of activity, so as to protect them from fatigue or stress, although no such requirement exists for operators. Should an employer fail to ensure that the above requirements are met, they may be subject to civil action in common law Negligence.

Regulation 5 requires the employer to provide eyesight tests for all users. These should be done at the request of a user, when being appointed as such, on a regular basis or whenever the user experiences visual difficulties. The employer also has to pay for any spectacles that may be required as a result of any deterioration in the user's eyesight.

Finally, Regulation 6 requires the employer to provide adequate health and safety training for users, before they start as a user and whenever the equipment is substantially modified. Under Regulation 7 there is a requirement to provide information on health and safety, both to users and operators.

Summary

The *DSE Regulations* have led to a need for employers to control the use of display screen equipment to avoid injury to employees, and also provide guidance on how to do this. While it is possible that technological advances may reduce the need for inputting data by keyboard in the future, the constant use of such equipment will still remain a problem for some time.

Physical hazards and hazardous substances

In this section I would like to look at those regulations that relate to the effect of physical hazards or the use of hazardous substances at work. In this respect I will be looking at seven regulations; namely, the *Electricity at Work Regulations 1989 (EAWR), Regulatory Reform (Fire Safety) Order*

2005, Control of Noise at Work Regulations 2005 (CNWR), Control of Vibration at Work Regulations 2005, (CVWR), Control of Substance Hazardous to Health Regulations 2002 (COSHH), Control of Asbestos Regulations 2012 (CAR) and the *Control of Lead at Work Regulations 2002 (CLWR).* These regulations cover a variety of physical and substance hazards in very different types of work activity and again they have generated some useful case law.

Electricity at Work Regulations 1989 (EAWR)

Scope of the regulations

The *EAWR* provide for safety in the use of electrical equipment. This is widely defined in Regulation 2 as including 'anything used . . . to generate, provide, transmit, transform, rectify, convert, conduct, distribute, control, store, measure or use electrical energy'. The duty is placed on the employer and self-employed persons, so far as it relates to matters within their control. The employee is under a duty to co-operate with his employer in order to ensure that he is able to carry out his duty.

Construction and maintenance of electrical systems

Under Regulation 4 all systems are to be of such construction as to prevent danger. Similarly, all systems must be maintained to ensure the same, and any work activity should be carried out in a manner to avoid danger. Under Regulation 7 all conductors in a system have to be suitably insulated or placed in such a position as will prevent danger. In all cases, the liability is qualified by the term 'reasonably practicable'.

Control measures

However, there are a large number of specific duties that are absolute in nature, except in as much as the defendant has a defence of showing that he took all reasonable steps and exercised all **due diligence**. The duties here include such matters as; ensuring the strength and capability of the electrical equipment (Regulation 5), providing suitable earthing (Regulation 8), providing for the integrity of the conductor (Regulation 9), providing suitable electrical connections (Regulation 10), protecting from excess current (Regulation 11), providing suitable means for cutting off the supply and isolating any electrical equipment (Regulation 12), taking adequate precautions for working on equipment made dead (Regulation 13) and ensuring the competence of the person working on any equipment (Regulation 16).

Work on live conductors

As regards working on live conductors, Regulation 14 limits this to situations where three factors apply: it is unreasonable for the equipment to be dead, it is reasonable for the equipment to be worked on live and where suitable precautions are taken. Again this is subject to taking all reasonable steps, as well as exercising due diligence, in order to avoid a breach of the law.

Summary

These regulations are comprehensive in nature and will apply to standard 240 volt supply systems as well as high voltage systems. However, the specialised activity of working on live equipment will require work to be carried out by persons who have a higher level of competence.

Regulatory Reform (Fire Safety) Order 2005

These regulations replace the fire certification system under the *Fire Precautions Act 1971* with a series of general duties, which are enforced by the relevant fire and rescue authority. They established a unified fire safety regime applicable to all workplaces and non-domestic premises, with responsibility for fire safety delegated to a responsible person at the relevant premises.

Scope of the regulations

The regulations apply to most workplaces and non-domestic premises, but this does not include locomotives or other vehicles, farmland or forestry operations.

Under Article 5 a responsible person is to be appointed who will ensure the carrying out of the duties imposed by Articles 8 to 22 of the order. Article 3 defines such a person as the employer or any person who has control of the premises, in respect of carrying on a business there.

General fire precautions

Under Article 8 the responsible person has to implement general fire precautions, as far as is reasonably practicable, to ensure the safety of the employees. These precautions will include; measures to reduce the risk and spread of fire, provision of means of escape, means of fighting the fire, as well as detecting and giving warning of such fires, and finally the provision of relevant instruction and training. In addition, he must implement such general fire precautions as may reasonably be required to ensure the safety of the premises.

Fire risk assessment

Under Article 9 the responsible person must make a 'suitable and sufficient' risk assessment to identify the controls required. This should be based on the matters set out in Schedule 1 of the order.

Control systems

Under Article 11 the responsible person must arrange for the effective planning, organisation, control, monitoring and review of the preventative and protective measures. Article 12 provides for the elimination or reduction of the risk from dangerous substances, these being defined as substances that are explosive, oxidising or flammable, as well as explosive dusts. Article 13 requires the responsible person to ensure that the premises are provided with appropriate fire-fighting equipment, as well as with smoke detectors and alarms. Article 14 requires the provision of emergency escape routes and exits and to ensure that such exits are kept clear at all times. Article 15 requires the establishment of procedures to deal with imminent danger and Article 16 the provision of additional emergency measures in respect of dangerous substances. Finally, Article 17 requires the responsible person to implement a suitable system of maintenance in respect of the above mentioned facilities, equipment and devices.

Information and training

Articles 19 to 22 require the provision of information and training for employees, as well as providing information to other employers working on the premises and implementing measures to ensure effective co-operation with such employers.

Duties on employees

Article 23 sets out the duties of employees and these are very similar to those in Regulation 14 of the *MHSWR*. In this respect, Article 23 not only requires the employee to co-operate with his employer, but also to bring to his attention any matter that could reasonably be seen as being a serious and immediate danger to safety, or that appears to be a shortcoming in the employer's protection arrangements.

Enforcement

As the Order is enforced by the fire and rescue authorities, the final articles provide for similar powers of inspection as the HSE has in other health and safety legislation, including the power to serve enforcement notices as well as to take prosecutions.

Summary

The structure of the order is very similar to the regulations enforced by the HSE, except that it deals with fire related issues. The role of the fire risk assessment is central here in identifying the control measures that need to be implemented to ensure the safety of the premises.

Control of Noise at Work Regulations 2005 (CNWR)

The *CNWR* replaces the earlier *Noise at Work Regulations 1989* and implements the EU Directive 2003/10/EC. The regulations came into force in 2006, but only apply to the music and entertainment industry from 2008.

Scope of the regulations

Regulation 2 defines noise as 'any audible sound', while the daily personal exposure is defined as 'the level of daily personal noise exposure of an employee as ascertained in accordance with Schedule 1 of the Regulations', and takes into account the level of noise and the duration of exposure. Exposure limit values are defined as the level of daily or weekly personal noise exposure or peak sound pressure that must not be exceeded, while upper and lower exposure action values are the two levels of daily or weekly personal noise exposure or peak sound pressure that, if exceeded, require specific action to be taken.

Duties of the employer

Regulation 3 places the duty on an employer in respect of ensuring the health and safety of his employees, so far as is reasonably practicable, in respect of any injury caused by their exposure to noise at work. This may also extend to other persons at work who may be so affected, with the exception of the duty to provide health surveillance and information. There is also a duty on a self-employed person, again with the exception of carrying out health surveillance on himself.

The exposure limits and action values are set out under Regulation 4. This defines a lower exposure action value limit of 80 decibels of A-weighted sound (or 80 dbA), and a peak sound pressure of 135 decibels of C-weighted sound (or 135 dbC). The upper exposure limits are set at 85 dbA and a peak sound pressure of 137 dbC respectively. In addition, there is a daily and weekly exposure limit value of 87dbA and a peak sound pressure of 140 dbA.

Noise assessment

Under Regulation 5 the employer is required to carry out a 'suitable and sufficient' assessment of the risk from noise if there is any possibility of exposing an employee to noise levels at or above the lower exposure action value. This would include the carrying out of noise monitoring if necessary. Any findings must be recorded, along with details of the measures to be taken to deal with the problem.

Regulation 6 sets out the methods that the employer should use to eliminate or reduce the risk to his employees. These are based on the general principles of prevention, as set out in Schedule 1 of the *Management of Health and Safety Regulations 1999*, and as such, include reducing exposure to noise through technical means, as well as by limiting the duration and intensity of exposure. In all events, the employer must ensure that his employees are not exposed to noise above an exposure limit value, or where that is the case, they must take action to prevent it being exceeded again.

Provision of and maintenance of personal protective equipment

Under Regulation 7 where an employee is likely to be exposed to noise at, or above, the lower exposure action value, then the employer is required to make personal hearing protection available. However, if the employee is likely to be exposed to noise at, or above, the upper exposure action value, then the employer must provide hearing protectors to such employees and ensure that they are worn. Moreover, such an area should be designated a Hearing Protection Zone and access to the area should be restricted where practicable. Regulation 8 provides for such hearing protection to be maintained in an efficient state.

Health surveillance, information and training

The employer has to provide health surveillance for his employees if there is a risk to health from exposure to noise (Regulation 9). In addition, there is a duty to provide relevant information, instruction and training (Regulation 10).

Civil actions

Most civil actions for noise induced hearing loss have been taken under common law Negligence as in *Berry* v *Stone Manganese and Marine Ltd (1972)*. Some cases have been brought under Section 29 of the *Factories Act 1961*, on the basis that a high noise level meant that the workplace was not 'safe' as required by the statute, as in the case of *Baker* v *Quantum Clothing PLC (2011)*. However, the most recent case of *Doyle* v *Advocate General for Scotland (2013)* involved a successful claim for breach of CNWR by a police weapons tester, although there was a reduction for contributory negligence on the part of the claimant.

It is likely that the claimant would still be able to win a case for breach of common law Negligence if he can prove a work-related noise induced hearing loss, where the employer has failed to take any action to reduce the risk. This reflects the decision in the similar case of *McCaffrey* v *Metropolitan Police District Receiver (1977)* involving a ballistics expert, where the court held that the damage caused was reasonably foreseeable.

Summary

These Regulations focus on the need for a noise assessment to see if the employees are exposed to noise levels in excess of the action levels. If so, the duty is on the employer to take action to control the noise at source, or failing that, to provide adequate personal protective equipment.

Control of Vibration at Work Regulations 2005 (CVWR)

Scope of the regulations

The *CVWR* implements EU Directive 2002/44/EC. It deals with the problem of hand-arm vibration, in which the use of vibrating tools is responsible for causing constrictions in the blood supply to the hands, leading to a condition called 'vibration white finger'. In addition, the regulation deals with whole-body vibration, which leads to other health problems such as nausea and loss of balance.

Regulation 2 defines hand-arm vibration as mechanical vibration that is transmitted into the hands and arms during work activity. Whole-body vibration is defined as mechanical vibration that is transmitted into the whole body while sitting or standing, through a supporting surface such as a seat. Mechanical vibration is defined as vibration occurring in a piece of machinery.

Under Regulation 3 the employer is under a duty to his employees, and in most cases he also has a duty to any person who may be affected by the work carried out, whether that person is at work or not. This is a very wide duty of care, qualified by what is reasonably practicable, and will extend to members of the public as well as persons at work. The employer has a duty to his employees only, in respect of arranging health surveillance under Regulation 7, while the duty to provide information, instruction or training applies to all persons on the premises where the work is being carried out.

Regulation 4 sets out the various control limits for hand-arm vibration and whole-body vibration. In the case of hand-arm vibration there is a daily exposure limit value of 5 metres per second squared (5 m/s^2) and a daily exposure action value of 2.5 metres per second squared (2.5 m/s^2). In the case of whole-body vibration, the respective values are 1.15 metres per second squared (1.15 m/s^2) and 0.5 metres per second squared (0.5 m/s^2).

Risk assessment

The employer has a duty to carry out a suitable and sufficient risk assessment under Regulation 5. This will look at matters such as the magnitude, type and duration of exposure, the effect of the exposure on the individual and existing equipment, as well as the availability of replacement equipment.

Control measures

After carrying out a risk assessment, under Regulation 6 the employer must then take action to control exposure to vibration at the workplace. Ideally, this will involve the elimination of the vibration, but if this is not possible, then it should be reduced to as low a level as reasonably practicable. In order to effect this, the employer should use a hierarchy of controls, starting with changes to working methods, continuing on to the use of low vibration equipment, the maintenance of work equipment, the limitation of the duration of exposure and eventually the provision of suitable protective clothing. Such controls must be implemented once an exposure action value has been exceeded, in order to reduce exposure to as low a level as reasonably practicable. However, if the employee is exposed to vibration above one of the exposure limit values, then the employer has to take action to prevent a recurrence.

Health surveillance, information and training

Under Regulation 7 the employer is required to provide health surveillance for employees, where they are exposed to vibration at, or above, an exposure action value, or if they are found to be suffering from a vibration-related disease. Such employees should be assigned to alternative work, where there is no risk of exposure to vibration. In addition, under Regulation 8 the employer has to provide relevant information, instruction and training for his employees if the risk assessment indicates a possible health risk for those exposed to vibration.

Summary

As with noise, the control over vibration is based on the employer assessing the level of vibration that the employee is exposed to and ensuring that the exposure limit values are not exceeded. Again, the need to eliminate the vibration at source is preferred, as opposed to providing personal protective equipment.

Control of Substances Hazardous to Health Regulations 2002 (COSHH)

The *COSHH Regulations*, as amended in 2004, are designed to protect workers against the risks posed by hazardous substances in the workplace.

The regulations are supported by an Approved Code of Practice and Guidance Note (L5), entitled 'Control of Substances Hazardous to Health'.

Scope of the regulations

Regulation 2 defines the term 'substance hazardous to health' as including those substances defined as very toxic, toxic, harmful, corrosive or irritant, as listed in the EU *Classification Labelling and Packaging Regulation (EC No. 1272/2008) (CLP)*. In addition, it includes any substance for which the HSE has approved a Workplace Exposure Limit (WEL), as set out in their Guidance Note (EH 40), entitled 'Workplace Exposure Limits'. In addition, it includes a substance that is a biological agent, as well as a dust of a certain concentration, or where it is a substance that is a risk to health as a result of its chemical or toxicological properties. It can be seen here that the definition is quite wide and it would appear that any substance can fall within the above definition, so it is the nature of the use of the substance that determines whether it is hazardous. Note that the regulations do not cover some hazardous substances that already have their own specialised control regime, such as asbestos and lead.

Duties on the employer

Regulation 3 specifies that a duty is placed on the employer in respect of any person, whether at work or not, who may be affected by the activities carried on by that employer. This is a much wider duty than that under *PUWER* and will cover a variety of third parties, including visitors and members of the public. This duty is limited to employees in respect of more specific matters such as carrying out health surveillance, environmental monitoring or providing training.

Risk assessment and control measures

Under Regulation 6 the employer has a duty to assess the risk to employees caused by the use of substances hazardous to health. Under Regulation 7 he has to take steps to prevent or control exposure of his employees to such substances. In this respect, he has to follow a hierarchy of control, seeking to eliminate or reduce the risk by substitution or changing the process. Only if this is not possible should he resort to control measures to limit exposure, such as enclosure or ventilation systems. While the duty to prevent exposure is qualified by the term 'reasonably practicable', the duty to control is absolute. This was made clear in the case of *Dugmore* v *Swansea NHS Trust (2002)*, where the claimant successfully obtained compensation for an anaphylactic attack, caused by exposure to dust from latex gloves.

Regulation 8 requires the employer to ensure that control measures are used by his employees, and Regulation 9 requires him to ensure that such controls are maintained, examined and tested so that they remain effective. An example here is a local exhaust ventilation system (LEV), which needs to be examined and tested every 14 months.

Monitoring of exposure

Under Regulation 10 the employer is under a duty to carry out monitoring of the exposure of employees to particular substances if there is a requirement arising from a suitable risk assessment. In addition, when appropriate, the employer should carry out health surveillance under Regulation 11 of *COSHH*. In both cases, records of personal exposure and personal medical records should be kept for a minimum of 40 years. This is so that they can be used by the employee, should he claim in respect of a latent work-related disease.

Training and emergency procedures

The other regulations under *COSHH* cover the need to provide information, instruction and training (Regulation 12), and to ensure that emergency arrangements are in place (Regulation 13). The regulations therefore provide a comprehensive system of controls and monitoring, to ensure the safety of employees in respect of hazardous substances at work.

Summary

The *COSHH Regulations* cover a wide variety of hazardous substances and focus on the need to assess the hazards and impose control measures. The importance of adequate training and emergency procedures is also key here. Some of the more hazardous substances, such as asbestos and lead, have their own specific legislation, both of which are detailed below.

Control of Asbestos Regulations 2012 (CAR)

The *CAR* amend the earlier regulations passed in 2006, which implemented the EU Directive 83/477/EEC on 'Asbestos Workers' Protection'. This area of health and safety is highly regulated as a consequence of our increasing understanding of the serious health effects of asbestos.

Scope of the regulations

Under Regulation 2 of *CAR* asbestos is defined as a series of fibrous silicates, including most commonly asbestos grunerite (or amosite), chrysotile or crocidolite. Work with asbestos is defined as involving the removal, repair or disturbance of asbestos, as well as any ancillary work or supervision of the above. Reference is made to the 'control limit', which is defined as a concentration of asbestos in the atmosphere of 0.1 fibres per cubic centimetre of air, averaged over a continuous period of four hours. Regulation 3 states that the regulations apply to employers, employees and self-employed persons.

Under Regulation 4 a duty to manage asbestos is placed on the 'duty-holder', this being the person who is under a contractual obligation to maintain or repair any non-domestic premises, or a person who has control over such premises where no contractual obligation exists. The dutyholder is required to assess whether asbestos is present or likely to be present in the premises, and if so, to determine the level of risk posed and to specify what measures are to be put into effect to control that risk. Those measures should include monitoring the condition of the asbestos, maintaining it or arranging for its removal, as well as providing information about the asbestos and its location to any person likely to disturb it.

Risk assessment

Regulation 5 requires an employer who undertakes any work that exposes employees to asbestos to carry out a suitable and sufficient assessment to identify the type of asbestos involved and its condition. Furthermore, under Regulation 6 this is to be followed by a suitable and sufficient assessment of the risks to health and the effect of any control measures taken, or any other steps to prevent or reduce exposure.

Planning of work operations

Regulation 7 requires the employer to prepare a plan of work, detailing how operations are to be carried out. In the case of the demolition or major refurbishment of premises, the plan must, so far as is reasonably practicable, specify the removal of the asbestos, unless such removal would pose a greater risk than leaving it in place. The plan of work must also provide details of the measures the employer intends to take in order to prevent or reduce exposure to asbestos.

Licensing system

Regulation 8 introduces a licensing system operated by the HSE, which covers 'licensable' work. This work is defined in Regulation 2 as work where exposure to asbestos is not sporadic and of low intensity or where it is not possible to show that the control limit will not be exceeded, as well as any work on asbestos coatings. The terms of such a licence and its period of operation can be varied by the HSE, or the licence as a whole can be revoked. In the case of licensable work Regulation 9 requires the employer to notify the appropriate enforcing authority where licensable work is to be carried out, as well as some non licensable work.

Information and training

Regulation 10 requires the employer to provide information, instruction and training for employees to cover a variety of matters. This would include the type of asbestos and its hazards, the operations that could result in asbestos exposure, the safe working practices to be adopted, relevant emergency procedures, decontamination procedures and procedures for disposing of contaminated waste.

Control measures

Under Regulation 11 the employer must prevent the exposure to asbestos of any employee, so far as is reasonably practicable. This involves taking measures to reduce exposure to the lowest level reasonably practicable, using a hierarchy of controls. These would begin with designing the work process to avoid or minimise the amounts of asbestos released, before continuing to controls such as adequate ventilation systems and eventually the provision of respiratory protective equipment. The aim is to ensure that no employee is exposed to asbestos above the level specified in the control limit.

Regulation 12 requires the employer to take reasonable steps to ensure that any control measures are properly used and employees are also required to do the same. The control measures must be 'maintained in an efficient state, in efficient working order and in good repair' by virtue of Regulation 13 and any local exhaust ventilation must undergo regular examinations. Under Regulation 14 the employer must provide adequate and suitable protective clothing for any person liable to be exposed to asbestos and such clothing must be cleaned at suitable intervals. Regulation 15 sets out arrangements to deal with accidents and emergencies, while Regulation 17 deals with ensuring the cleanliness of the premises and plant. Note here that the duties under Regulations 13, 14, 15 and 17 are all strict liability ones.

Zoning and personal protective equipment

Under Regulation 18 the employer is required to designate certain areas as asbestos or respirator zones. An asbestos zone is an area where an employee is likely to be exposed to asbestos, and entry to such an area must be restricted to persons required to be there for work purposes. A respirator zone is an area where the exposure of an employee is likely to exceed the control limit. Only competent persons must be allowed to enter such an area, and they must wear appropriate respiratory protective equipment.

Environmental monitoring and health surveillance

Regulation 19 places an obligation on the employer to monitor the exposure to asbestos of any employee, both at regular intervals and when a change occurs that may affect the exposure levels. Similarly, under Regulation 22 every employer must maintain a health record for employees so exposed, as well as a record of any medical surveillance. The above information in Regulations 22 must be retained for at least 40 years.

Welfare facilities and disposal of waste

Finally, an employer must provide washing and changing facilities under Regulation 23, and effective storage and disposal arrangement for waste material containing asbestos under Regulation 24.

Summary

The asbestos regulations are very comprehensive and cover a wide variety of matters. Recent cases have mainly been brought under the older regulations or in common law Negligence, and with the end of civil actions for breach of statutory duty it is unlikely that the above regulations will be actually tested in the courts in a civil action. However, it is possible that the present statutory requirements may be used as a basis for determining what controls would be expected of a reasonable employer when establishing liability in Negligence.

Control of Lead at Work Regulations 2002 (CLWR)

The *CLWR* implement a number of EU directives, such as 82/605/EEC, 80/117/EEC and 98/24/EC. The regulations are accompanied by an Approved Code of Practice as well as a Guidance Note (L132), entitled 'Control of Lead at Work'.

Scope of the regulations

Under Regulation 2 lead is defined as lead (including lead alkyls, lead alloys and any compounds of lead) that is liable to be inhaled, ingested or otherwise absorbed by persons, excluding lead in vehicle exhaust fumes. Regulation 2 also defines what constitutes the relevant 'action level', which is different for persons of reproductive capacity, young persons or other employees. The occupational exposure limit is defined as 0.15 milligrams per cubic metre (0.15 mg/ m^3) for lead and 0.1 milligrams per cubic metre (0.1 mg/ m^3) for lead alkyls, in relation to an eight hour time-weighted average. The regulations define what constitutes the relevant 'suspension level', at which an employee must be suspended from work on health grounds, as well as setting out what constitutes 'significant' exposure to lead.

Under Regulation 3 where a duty under the regulations is placed on an employer in respect of his own employees, it also applies in respect of any other person, whether at work or not, who may be affected by the activities carried out by the employer. This is a wider definition than most subordinate regulations but is qualified by the term 'reasonably practicable'.

Risk assessment

Under Regulation 5 an employer is required to carry out a suitable and sufficient risk assessment of work that is liable to expose any employee to lead. Such an assessment should look at the hazards posed by the lead, the nature of the exposure, the relevant occupational exposure limit, action level and suspension level, as well as the effect of any preventative and control measures. In addition, the employer must consider whether the exposure of any employee is 'significant', as defined by Regulation 2.

Control measures

Regulation 6 places a duty on the employer to prevent or control exposure of his employees to lead. This involves the use of a hierarchy of controls, including substitution, process controls, ventilation and the use of personal protective equipment. He is also required to make arrangements for the safe handling, storage and transport of lead and lead waste, as well as ensuring adequate hygiene measures. The aim is to ensure that the occupational exposure limit is not exceeded, or where it is, that immediate steps are taken to remedy the situation.

In order to ensure that exposure to lead is effectively controlled there are some extra requirements set out in the rest of the regulations. Therefore Regulation 7 requires the employer to ensure, so far as is reasonably practicable, that there is no eating, drinking or smoking in an area liable to be contaminated by lead. In addition, Regulation 8 requires the employer to examine and test the relevant control measures, including local exhaust ventilation systems.

Environmental monitoring and medical surveillance

Regulation 9 requires the employer to carry out air monitoring when employees are likely to receive significant exposure to lead. Regulation 10 sets out a comprehensive arrangement for carrying out medical surveillance. This is required where employees are subject to significant exposure to lead, where a doctor certifies that an employee should be under such surveillance or where the blood-lead levels or urinary-lead levels exceed certain limits. In addition, where the blood-lead level exceeds the appropriate action level the employer is required to take steps to determine the reason for this and to introduce measures to tackle the problem. In some cases, this may lead to employees being suspended from any work involving exposure to lead or being employed under certain restrictions.

Information and emergency procedures

Apart from the above, there is a requirement to provide relevant information, instruction and training under Regulation 11, as well as providing arrangements to deal with accidents, incidents and emergencies under Regulation 12.

Summary

The serious hazards posed by the use of lead and lead compounds at work is the reason for these specific regulations. The combination of risk assessment, control measures and monitoring systems provides a comprehensive system of controls to provide effective protection for employees.

Conclusion

The above regulations are varied in nature but have many common elements. They all involve the use of risk assessments, control measures and monitoring systems. However, in some cases, the employer needs to provide environmental monitoring, as well as medical surveillance, to ensure effective protection of employees. While the above section does not cover all of the subordinate regulations, it does deal with the key ones that the employer is likely to meet and gives a flavour of the key elements involved.

Notes

1   'Five steps to risk assessment' *INDG* 163 (Rev 2) by the Health and Safety Executive (HSE)
2   OHSAS 18001:2007
3   Nigel Tomkins 'The workplace regulations and strict liability' *Journal of Personal Injury Law* 2010, 65–69
4   See HSG 38 'Lighting at work' by the HSE
5   See p. L3036 *Tolley's health and safety work handbook* 2010
6   23 May 2013 QBD unreported, see Nigel Tomkins 'Case comment' *Journal of Personal Injury Law* 2013, 2, C95–98

6 Enforcement of criminal liability

Introduction

Criminal liability in health and safety law is enforced in a number of ways, both direct and indirect. The main form of direct enforcement is by means of statutory enforcement bodies, such as the Health and Safety Executive (HSE) or the local authorities. There are a number of other relevant enforcement bodies such as the fire and rescue authorities or the Rail Regulatory Authority, but their role is generally peripheral, so I will concentrate on the two main organisations.

In addition, there is a form of indirect enforcement, which is carried out by insurance companies and professional organisations. An insurance company will refuse to insure or reinsure an organisation if they feel that its health and safety management system is inadequate, because otherwise they may incur substantial costs if they have to pay out compensation for an accident. On the other hand, a professional organisation may threaten to withdraw accreditation unless an organisation improves its health and safety performance, and it is usually a benefit for an organisation to obtain safety accreditation, such as OHSAS 18001. For many employers these are very effective forms of enforcement, albeit indirect in nature. Finally, we should note here the role of the safety representatives, who often exercise some element of influence over health and safety matters, especially in unionised workplaces, although they are not an actual enforcement organisation.

Statutory enforcement

As regards statutory enforcement, both the HSE and the local authorities enforce the *HSWA* and the subordinate legislation passed under it. The two organisations divide their responsibilities under the provisions of the *Health and Safety (Enforcing Authority) Regulations 1998*. In general, the HSE will undertake the inspection of factories, fairgrounds, quarries and mines, while the local authority will inspect offices, retail outlets and warehouses. However, the HSE has always taken a role in directing the enforcement policies of the local authorities in order to ensure a consistent approach, and there are renewed suggestions that HSE will take overall control of the local authorities as regards their enforcement role in health and safety. In this section I will look at the powers, as well as the liabilities of inspectors, the role of prosecution in enforcement and the enforcement notice system.

The powers of inspectors

The role of enforcement falls to the Health and Safety Inspector, either in HSE or as an Environmental Health Officer in a local authority. In both cases the inspector will have the same powers as set out under Section 20 of *HSWA*, which are very extensive in nature. However, the role of an inspector is not just to enforce the law, but also to give advice, and sometimes it will be the case that there is a tension between the two roles. In addition, the recent introduction of charging for inspection work, under the Fee for Intervention policy, adds another complicating factor, which I will refer to later.

Power of entry

Under Section 20 of *HSWA* an inspector has a variety of powers, but he must produce his warrant if requested in order to enforce them. He can enter premises at any reasonable time or at any time if he thinks the situation is dangerous. If he is prevented from entering, he can take a police constable with him in order to effect an entry. The inspector can also take another person authorised by the enforcing authority into the premises, such as a specialist inspector, as well as any equipment or materials the inspector may require.

Power of obtaining information

Once in the premises, the inspector has various powers under Section 20 of *HSWA*, which he can use to obtain information necessary for him to perform his job. The inspector can carry out any examination or investigation that he considers necessary. Moreover, he can direct that the premises or any part of them are left undisturbed, so long as it is necessary to carry out any examination or investigation. This is important so that vital evidence is not compromised and because he may need to call in a specialist to carry out such an examination at a later time.

The inspector can take measurements, photographs and any recordings as he thinks necessary. He can also take samples of any article or substance found in the premises. Moreover, he can require any article or substance that appears to him to be likely to cause danger to be dismantled or subject to any test, and he can take possession of such an article or substance and retain it in order to carry out this testing.

The inspector can ask such questions as he feels are appropriate, of any person whom he believes can provide information relevant to his examination or investigation. Unlike police questioning under the *Police and Criminal Evidence Act 1984 (PACE)* that person is obliged to provide an answer to the questions asked and to sign a declaration of truth. However, such answers cannot be used as evidence against that person in any legal proceedings, so the inspector would have to carry out a separate interview under the requirements of Section 67 of *PACE*.

An inspector can take copies of any entries in books or documents that are necessary for the inspector to see to enable him to carry out his duties. However, this will not cover documents that are protected by legal professional privilege, such as an accident investigation report prepared for the employer's legal representative, as in *Waugh v British Railways Board (1980)*.

Power to deal with imminent danger

In addition to his powers under Section 20, there is a separate power under Section 25 of *HSWA* to deal with causes of imminent danger. The inspector can seize any article or substance that he has reason to believe could cause imminent danger and require it to be rendered harmless. If he exercises this power, he should, if practicable, give a sample of the article or substance to the employer, as well as a written report giving the particulars of the circumstances in which the article or substance was seized. This enables the employer to have the article or substance tested in the case of any subsequent legal action that may take place.

Obtaining information

In addition to the powers outlined above, an inspector also has the power to obtain information under Section 27 of *HSWA*, where this is required by the enforcing authority. This is done by serving a notice on the relevant person, requiring him to provide the information as specified in the notice. Failing to comply with such a notice will render that person subject to a fine not exceeding the statutory maximum in the Magistrates' Court, or an unlimited fine in the Crown Court.

Enforcement action

Having carried out an examination or investigation there are various enforcement mechanisms available to the inspector if required. He may decide to take a prosecution against the employer or any individual who is in breach of any of the relevant statutory provisions. The inspector can also serve an improvement or prohibition notice, by virtue of Sections 21 and 22 of *HSWA* respectively, as detailed below. He can also write a letter to the employer detailing which actions should be taken to improve matters, with the possibility of taking further action if there is no change in the situation.

The above powers are very extensive and if an employer tries to obstruct an inspector he commits an offence under Section 33(h) of *HSWA*. However, if the inspector exceeds his power, he may find that the court will strike down any legal action brought by him against an employer, as was the case in *Skinner* v *John McGregor (Contractors) Ltd 1977*.

Liability of inspectors

Under the provisions of section 28 of *HSWA* any information obtained by the inspector can only be used for specific purposes, unless the parties giving that information consent otherwise. The information can only be disclosed to a designated statutory authority (such as the Environment Agency), used for the purposes of legal proceedings or to provide information to employees or their representatives, in respect of matters affecting their health, safety and welfare. Confidential information may be disclosed, if necessary, to prevent a criminal act, as referred to in the case of *R (on the application of TB)* v *Stafford Crown Court (2006)*.[1]

An inspector who divulges information in breach of Section 28 may be personally liable for his actions if it causes economic loss to the employer. Similarly an inspector can be personally liable if his actions cause economic loss as a result of negligently given advice, although the relevant enforcing authority may actually take responsibility. An example here is the case of *Welton* v *North Cornwall District Council (1997)*, where the local authority Environmental Health Officer required extensive kitchen refurbishment to be carried out in a hotel, which was later deemed to be unnecessary. The local authority was held responsible for the refurbishment costs incurred by the hotelier.

However, the court will not necessarily penalise an inspector simply because his decision is later questioned by an Employment Tribunal, as this would hamper inspectors in the execution of their duties. This can be seen in the case of *Harris* v *Evans and the Health and Safety Executive (1998)*, where the inspector banned the use of a crane for bungee jumping, but the decision was quashed on appeal to the Employment Tribunal. The bungee jump operator was unable to claim the loss of profit from being unable to carry out his activities in the intervening period.

Prosecution of offences

Power of prosecution

An inspector can prosecute an employer, an individual employee or any other person if there is a breach of a relevant statutory duty. Section 33(1) of *HSWA 1974* sets out the specific offences that a person may be charged with, most of which are either way offences but some of which are summary. The present position is that Sections 2–9 of *HSWA* are either way offences, as are the health and safety regulations passed under the act. In addition, breaches of Sections 20, 25 and 28 are also either way offences, while obstructing an inspector or falsely pretending to be an inspector are summary offences only. Offences under the *Corporate Manslaughter and Corporate Homicide Act 2007* are all indictable and so go to the Crown Court, as do any individuals prosecuted for common law Manslaughter.

Sentencing powers

The sentencing powers of the court are set down in Section 33 of *HSWA* and these have been recently modified by the *Health and Safety (Offences) Act 2008 (HSOA)*. This now allows for custodial sentences in both the Magistrates' Court and the Crown Court, for breaches of *HSWA* or any subordinate legislation passed under the act. In the Magistrates' Court the maximum penalty is a fine of £20,000 or six months imprisonment, while in the Crown Court it is an unlimited fine and up to two years imprisonment. In addition, it is possible for a director of a company to be disqualified under the provisions of the *Company Directors Disqualification Act 1986*, and for the holder of a licence, such as an asbestos stripping licence, to have that licence revoked. The key levels of penalty under *HSOA* are as set out in Table 6.1.

Prosecution policy

In deciding whether to prosecute, inspectors are expected to follow the guidance provided by the HSE in their Enforcement Policy Statement.[2] This sets down the criteria which are used to decide whether or not to prosecute. The factors favouring a prosecution include; whether there has been a death, the gravity of the offence, the seriousness of the harm occurring, whether there has been reckless disregard of health and safety, the failure of the employer to comply with the law in the past and whether inspectors have been intentionally obstructed. In England and Wales the inspector takes the case in the Magistrates' Court, but a barrister takes the case on his behalf in the Crown Court. In Scotland the Procurator Fiscal decides whether to take the case in the Sheriff's Court, sometimes acting on the recommendation of the HSE. If the HSE decide not to take a prosecution the decision is subject to judicial review, as per *R v DPP ex part Timothy Jones (2000)*, as discussed previously.

In the past five years the number of prosecutions has been fairly steady, with the HSE and local authorities together bringing between 1,200 and 1,400 cases a year as set out in Table 6.2.[3] The conviction rate has also remained consistent at around 85 per cent of cases brought. However, the number of prosecutions is well down on the figure at the beginning of the decade, when there were nearly 2,200 cases, although with a lower conviction rate.[4]

Table 6.1 Levels of penalty

Offence	Summary conviction	Conviction on indictment
Breach of Sections 2–6 of *HSWA*	£20,000 fine and/or 6 months imprisonment	Unlimited fine and/or 2 years imprisonment
Breach of Section 7 of *HSWA*	Statutory maximum fine and/or 6 months imprisonment	Unlimited fine and/or 2 years imprisonment
Breach of Section 8 of *HSWA*	£20,000 fine and/or 6 months imprisonment	Unlimited fine and/or 2 years imprisonment
Breach of Section 9 of *HSWA*	£20,000 fine	Unlimited fine
Breach of Health and Safety Regulations	£20,000 fine and/or 6 months imprisonment	Unlimited fine and/or 2 years imprisonment
Obstructing a person exercising powers of investigation under Section 14 of *HSWA*	Fine up to level 5 on the standards scale	
Contravening the powers of an inspector under Sections 20 and 25 of *HSWA*	£20,000 fine and/or 6 months imprisonment	Unlimited fine and/or 2 years imprisonment
Contravening a requirement of an improvement or prohibition notice	£20,000 fine and/or 6 months imprisonment	Unlimited fine and/or 2 years imprisonment
Intentionally obstructing an inspector	Fine up to level 5 on the standard scale and/or 6 months imprisonment	
Failure to provide information under Section 27 of *HSWA*	Fine not exceeding the statutory maximum	Unlimited fine
Using or disclosing information in contravention of Sections 27 or 28 of *HSWA*	Statutory maximum fine and/or 6 months imprisonment	Unlimited fine and/or 2 years imprisonment
Making a false statement or a false entry in a register or other document or using a forged document	£20,000 fine and/or 6 months imprisonment	Unlimited fine and/or 2 years imprisonment
Falsely pretending to be an inspector	Fine up to level 5 on the standards scale	
Failing to comply with an order of court	£20,000 fine and/or 6 months imprisonment	Unlimited fine and/or 2 years imprisonment

Table 6.2 Number of prosecutions, convictions and level of fine (HSE and Local
        Authorities for Great Britain)

Period	Prosecutions	Convictions	Conviction rate (%)	Average fine (£)
2008/9	1,373	1,153	84	12,147
2009/10	1,169	984	84	13,581
2010/11	1,206	1,041	86	20,172
2011/12	1,190	972	82	18,469
2012/13 (provisional)	1,217	1,069	88	13,995

See here www.hse.gov.uk/STATISTICS/tables/index.htm, go to table EF3 (accessed 20 August
2014)

The level of fines

In the 1980s there was some concern about the low level of fines being
imposed for health and safety offences. As a result, the maximum fine in the
Magistrates' Court was raised to £20,000 under the *Offshore Safety Act
1992* and the level of fines started to rise. Some guidance was given by the
Court of Appeal in the leading case of R v *Howe & Sons (Engineers) Ltd
(1999)*, which laid down various criteria for determining the level of the fine.
Aggravating factors that would lead to a higher fine included; how far the
employer fell short of the appropriate standards, whether there was a fatality
and whether there was a deliberate breach of the law. On the other hand,
the factors that would mitigate the level of a fine included; a prompt
admission of guilt, immediate steps to remedy the problem and a good safety
record. Up to a third can be deducted from a fine when there is a prompt
admission of guilt, as made clear in R v *Caley (David) & Others (2012)*.[5]
Indeed, following the case of R v *Friskies Petcare (UK) (2000)* the prosecution
and defences are now required to specify the relevant aggravating and
mitigating factors involved.

The result has been a gradual increase in the level of fines, with some very
high fines imposed on large companies. This includes a £10 million fine on
Balfour Beatty for the Hatfield crash and a record £15 million imposed on
Transco in 2005, following a gas explosion in Larkhill in Scotland. However,
as we can see in Table 6.2 the average level of fine in still quite low, at around
£16,000 per conviction over the last five years, and that figure is boosted by
a few large fines. As regards Corporate Manslaughter, it had been suggested
that fines should start at £500,000, yet so far there have only been a handful
of prosecutions, mainly against small and medium sized companies, and the
maximum fine imposed is still less than this figure.[6] In addition, it should
be noted that an organisation which is successfully convicted may have to
meet some of the investigation costs of the enforcement authority, following

the decision in the case of *R* v *Associated Octel (1995)*. This is further supplemented by extra costs imposed under the Fee for Intervention system, as detailed below.

Imprisonment

In addition to a fine, it has always been possible to imprison people for some health and safety offences, especially for refusing to comply with an enforcement notice, and of course for manslaughter. Following the passing of *HSOA* this has been extended to most offences under *HSWA*, but the number of persons actually imprisoned each year is very low, rarely exceeding double figures. In reality, only directors of small companies are likely to be imprisoned and then usually following a death at work. Examples here include the imprisonment for 14 years of the gangmaster in charge of the Chinese cockle pickers who drowned in Morecambe Bay, see here *R* v *Lin Liang Ren (2000)*, and the two directors imprisoned following the fatal accident caused by a runaway rail truck at Tebay in Cumbria in 2004.

Disqualification of directors

The ability to disqualify directors under the *Company Directors Disqualification Act 1986* is a power that has rarely been used, with only seven directors disqualified over the whole period of 1986 to 2005 and three in 2010–2011.[7] The lack of activity here was reported in a HSE research report from 2007.[8] It is arguable that this is a sanction of limited value, as a disqualified director can still operate through a substitute director, such as a friend or relative.

Other than the above, the type of punishment imposed has been rather limited in scope. It is possible for a judge to order a company to carry out remedial work. It is also the case that the poor publicity is a punishment in itself and will lead indirectly to a loss of business. However the HSE has put forward suggestions for alternative penalties in a recent consultative document.[9] These include such ideas as the use of restorative justice, fixed penalty fines, remedial orders, adverse publicity orders and placing directors on probation. There is a great deal of interest in how such ideas can be developed, but as yet very little action.

Enforcement notices

Introduction

In addition to a prosecution, it is possible for the HSE to use the enforcement notice system. This is a form of **quasi-legal** liability, because the notice serves as a warning to the employer to comply with the inspector's request, failure to do so leading to a prosecution. There are two main types of enforcement

Table 6.3 Number of enforcement notices served

Period	Improvement	Prohibition	Total
*(a) (HSE and Local Authorities)**			
2008/9	9,755	4,662	14,417
2009/10	10,474	5,363	15,837
2010/11	12,747	5,561	18,308
2011/12	10,750	5,203	15,953
2012/13	9,115	4,388	13,503
*(b) HSE only***			
1994/5	6,512	4,296	10,808
2003/4	6,798	4,537	11,335
2012/13	5,757	3,053	8,810

*   See here www.hse.gov.uk/STATISTICS/tables/index.htm (accessed 20 August 2014), go to table EF6
** HSE Statistics 2005/6 and Table EF6

notice used by the HSE, the improvement notice and the prohibition notice. Both are different in respect of the rules for serving them, the effect of the notice and the procedure for appealing against them.

The number of enforcement notices served in recent years has been constant at about 15,000 a year, although there is evidence of a small downturn in the numbers in 2012–2013. About a third of the notices are prohibition notices, and slightly more are served by the HSE rather than the local authorities (see Table 6.3).[10]

Improvement notices

Under Section 21 of *HSWA* an inspector can serve an improvement notice on an employer where he believes there is a breach of a relevant statutory provision. Such a notice gives the employer time to remedy the situation, at least 21 days and often up to two months. The work to be done is usually specified in a schedule appended to the notice.

Appeal against an improvement notice must be made within 21 days to an Employment Tribunal. The tribunal can uphold the notice, overturn it or vary its contents. While the appeal is waiting to be heard the improvement notice will be suspended. The appellant has to discharge the burden of proof in such a hearing, but the standard of proof is the civil one of the balance of probabilities.

An improvement notice may be overturned for a number of reasons. The tribunal may find that there is no actual breach of the law, as was the case in *Davis & Sons* v *Leeds City Council (1976)*, where the employer was able to show that he complied with the statutory requirement for provision of sanitary facilities, by using the facilities provided by adjacent premises. It is also possible to argue that the requirements imposed by the notice were not 'reasonably practicable' in the circumstances. Therfore in *Associated Dairies* v *Hartley (1979)* the employer successfully appealed against an improvement notice that would have required him to pay for safety boots for all of his employees, even though only a small number of them were at the risk of injury to their feet.

On the other hand, it is not possible to appeal against a notice on the grounds that the equipment had been used for a long time without an accident occurring. This point was raised in *Murray* v *Gadbury (1979)* in respect of a piece of farming machinery that had been used for a long time in an unguarded state. Despite the arguments of the employer, the tribunal still enforced the notice. It is no defence to claim that the required improvements impose an unreasonable financial burden on the business, but it is possible that the tribunal will delay the introduction of the notice to allow a company to finish an urgent order.

An improvement notice is a valuable means of obtaining compliance with health and safety law. Although not in itself a criminal conviction, it is clear to the employer that he will have to comply or he will be subject to a prosecution in circumstances where it is a simple matter for the HSE to win their case. However, with the introduction of Fee for Intervention, the service of a notice now comes with a charge of £1,500, which makes it more like a fixed penalty punishment. As a result it is possible that we will see more challenges to improvement notices in the future.

Prohibition notices

Prohibition notices may be served by an Inspector under Section 22 of *HSWA* where he believes there is a risk of serious personal injury. The inspector may also refer to a breach of a relevant statutory provision, although this is not strictly necessary. As with an improvement notice, a prohibition notice may give details of how to remedy the situation in a schedule attached to the notice. Note that a prohibition notice can be served even when there is no immediate possibility of the activity being re-started in order to ensure that certain actions are carried out, see the case of *Railtrack PLC* v *Smallwood (2001)*.

When an appeal is made against a prohibition notice, the notice still remains in force, unlike an improvement notice. Given the problems this may cause to an employer it is possible to obtain a hearing for an appeal against a prohibition notice the next day, as was the case in *Hoover* v *Mallon (1978)*. The onus is on the appellant to refute the prima facie presumption that the situation raised a risk of serious injury, but the standard required is that of Civil Law, in other words, the balance of probabilities.

The tribunal is reluctant to overturn a prohibition notice and it will not strike it down simply because it imposes too great a financial burden on the appellant, otherwise this would justify ignoring health and safety on the grounds of cost, see here the decision in *Nico Manufacturing* v *Hendry (1975)*. It is possible for a notice to be challenged if there is no evidence of a previous accident on a particular machine, as in the case of *Brewer & Sons* v *Dunston (1978)*. This would suggest that the risk of serious injury had been exaggerated, but note here the decision in *Murray* v *Gadbury (1979)*, as mentioned above.

A prohibition notice is clearly a greater imposition on an employer because of its immediate effect and it is not surprising that there may be pressure put on the inspector to defer its operation. Indeed, it is possible to serve a deferred prohibition notice under Section 23 of *HSWA* and such an order may even be appealed on the grounds that the time delay is too short. This was the case in *Otterburn Mill Ltd* v *Bulman (1975)*, where the tribunal refused to extend the time delay of three months on the basis that it would be unfair to allow an extension of time for purely commercial reasons. However, the fact that there had been no accidents for nine years on the machines involved meant that the tribunal was prepared to allow an extension of time on one of the machines.

The use of deferred prohibition notices does pose a real problem if a person is injured in the intervening period, before the prohibition notice was due to come into effect. This may expose the HSE to a civil action by the injured party, and it may be the reason why such notices are rarely used.

Other enforcement methods

Statutory regulation systems

In addition to the above methods of enforcement, inspectors have recourse to other regulatory systems that allow the enforcement authorities to exercise much higher levels of control and will require the relevant organisation to demonstrate its competency to operate. These can be divided into various types as set out below.

Regulatory review

Organisations that operate a hazardous activity, such as chemical plants or nuclear power installations, may be subject to extra regulatory systems. Under the *Control of Major Accident Hazards Regulations 1999 (COMAH)* the operator has to produce a Safety Report, which is sent to the enforcement authority, who in turn will review its contents and take action if it is not adequate.

Registration

In other cases, an operator has to be registered with the enforcement authorities who will then monitor his activities and carry out inspections during a probationary period. Failure to comply with the required standard will mean that registration may be withdrawn. An example here is a person working on gas appliances, who is regulated under the *Gas Safety (Installation and Use) Regulations 1998*.

Licensing

In this system the regulator grants a licence for the undertaking of a specific activity. The licence can then be withdrawn if the operator is found to be acting unsafely. An example here is the licensing of asbestos stripping organisations under the *Asbestos (Licensing Operations) Regulations 1983*.

Informal regulation systems

There are two main types of regulatory system that can exercise some form of control over a business organisation in respect of health and safety issues. The first is external, and involves the influence exercised by commercial organisations such as insurance companies and accreditation organisations. The second is internal, and involves the pressures exerted by employees, especially safety representatives.

External commercial controls

In addition to the regulatory system set up by the state, we should not forget the independent system that operates on a commercial basis. An example here is the role of the insurance companies, who will refuse to insure organisations seen as a poor safety risk. In some cases the insurance company will require evidence of a good health and safety management system and they may send in their employees to verify that such a system exists. The failure to arrange employee liability insurance means that the organisation is operating illegally, so the insurance company is able to exercise real control over its health and safety management systems.

In addition, many organisations seek accreditation from bodies such as the British Standards Institute who offer the OHSAS 18001 award, as this will enable them to obtain business more easily. The accrediting body can ensure that the relevant organisation maintains a satisfactory standard of health and safety by threatening to withdraw that accreditation.

Finally, we could note that clients will often refuse to engage an organisation that has a poor health and safety record to carry out work on their behalf. This is because they themselves may incur a liability under Section 3 of *HSWA* if they engage a party who is unable to properly manage its health and safety. This can be seen in the case of *R v Associated Octel (1996)*, where the operators of a chemical plant were held to be prima facie liable under *HSWA* because of the failure of a subcontractor to follow a permit to work system when carrying out repairs to a tank lining. As a result of this, one of the contractor's employees was injured. It is for this reason that a client will ask for detailed health and safety information before engaging a contractor to work on its premises. As such, this effectively operates as another informal method of enforcing health and safety legislation.

The role of safety representatives

There are two types of safety representatives operating in the workplace. The first are the Trade Union appointed safety representatives, regulated by the *Safety Representatives and Safety Committee Regulations 1977 (SRSCR)*, which is supported by two Codes of Practice. The second are the representatives of employee safety, appointed by the workforce under the auspices of the *Health and Safety (Consultation with Employees) Regulations 1996 (HS(CWE)R)*.

TRADE UNION SAFETY REPRESENTATIVES

Under Regulation 3 of *SRSCR* trade union appointed safety representatives are appointed by a trade union, which is independent of management control and as such is certified under Section 2 of the *Trade Union and Labour Relations (Consolidation) Act 1992*. Under Section 2(6) of *HSWA* it is the duty of every employer to consult such representatives in respect of health and safety matters at work, and under Section 2(6) of *HSWA* the employer must set up a safety committee if requested to do so by two of more such representatives. As far as possible, such persons should have been employed in the relevant workplace or a similar environment for at least two years.

A trade union safety representative has various rights, as set out in Regulation 4 of *SRSCR*. He has the right to investigate hazards and dangerous occurrences, to carry out workplace inspections, to investigate complaints from employees and to represent employees in discussions with management. Although a safety representative has no formal powers to require an employer to comply with the law, he can bring his concerns to the attention of the relevant enforcing authority. Moreover, as we have seen in Chapter 4, he has protection against unfair dismissal for carrying out his duties as a safety representative, by virtue of Section 44 and 100 of the *Employment Rights Act 1996*. In addition, he has the right to time off for carrying out the role, as well as time off for suitable training, which may be provided by the union or by the employer, as made clear in *White* v *Pressed Steel Fisher Ltd (1980)*.

The key right here is the right to carry out an inspection, as set out in Regulation 5 of *SRSCR*. This allows a safety representative to inspect the workplace at least once every three months and more frequently if there are any new developments or changes. As such, a safety representative may carry out an inspection following an accident or a dangerous occurrence and this may even involve him inspecting premises outside his place of work, as in the case of *Dowsett* v *Ford Motor Corporation (1981)*. The right of inspection also allows the safety representative to inspect relevant documents and take copies of the same, in order to enable him to perform his functions. In addition, the safety representative can also investigate any accident that occurs at the workplace and this may also involve carrying out a further inspection.

The rights of the safety representative under the *SRSCR* are quite extensive and provide him with a means of exercising real influence over health and safety matters. The safety representative is to some extent a relic of the concept of corporatism, which flourished in the 1970s and promoted employee participation in the running of industry. While the idea effectively disappeared in the 1980s for most purposes, it still continues to function in the area of health and safety in the role of the safety representative.

As an employee, a safety representative is bound by the normal duties of an employee under Sections 7 and 8 of *HSWA*. His role as a safety representative does not impose any extra duty upon him and he is not acting as a part of management. As such, he is in an unusual situation, with certain rights akin to those of management, but without any corresponding legal liability for his actions. However, a safety representative must always operate through the normal internal procedures of his employer where possible; otherwise he may be seen as acting in breach of his contract of employment.

A safety representative can contact the relevant enforcement authority in respect of any health and safety concerns and will be protected from disciplinary action not only under the provisions of the *Employment Rights Act 1996*, but also under the *Public Interest Disclosure Act 1998*. He is also entitled to speak to an inspector in confidence and can receive information from an inspector under Section 28(8) of *HSWA*, where it is related to matters affecting the health and safety of the workforce. As a last resort, a trade union safety representative can also use the threat of industrial action, if necessary, to help enforce health and safety law.

REPRESENTATIVES OF EMPLOYEE SAFETY

Representatives of employee safety are elected by the workforce and have certain rights under *HS(CWE)R*. These include the right to make representations to the employer and to be consulted on health and safety matters. However, unlike the trade union safety representatives, they do not have the right to investigate accidents or to carry out workplace inspections, although they may do this with the permission of the employer. As such, their influence is obviously much less, but employers may be reluctant to sideline them completely, as this could undermine the morale of their workforce.

Instead of dealing with representatives of employee safety, the employer may consult directly with the workforce. Under *HS(CWE)R* this should be done in respect of any measure that may substantially affect the health and safety of the workforce, including the appointment of 'competent persons' with key safety management functions, as well as the introduction of new technology. Such consultation allows employees to exercise a limited amount of influence over health and safety matters, in as much as the employer may wish to modify arrangements in order to maintain the goodwill of his employees.

Fee for intervention

However, one recent complicating factor is the introduction of the Fee for Intervention system introduced in October 2012. This enables the HSE to recover the cost of an inspector's time at a rate of £124 an hour. Such a fee must be recovered where the inspector determines that there has been a material breach of health and safety law. This means that the employer is charged when enforcement notices are served, but in addition he may also be billed when receiving a letter. This could mean that Inspectors become more reluctant to send letters detailing improvements to be made, but will rely on oral advice. As yet it is too soon to see what the long term implications are of Fee for Intervention, but the subject has generated a lot of controversy.[11]

Conclusion

We can see that organisations are subject to a wide measure of enforcement mechanisms that help to ensure compliance with health and safety laws. Some of them are formal systems, while others rely on informal pressure. We should remember that inspectors will not usually serve enforcement notices or prosecute, but will rely on persuasion to achieve results. In this respect, the sending of a letter requiring specific action will usually be effective in ensuring compliance.

Notes

1 See the reference on the HSE website www.hse.gov.uk/enforcement/guide/investigation/physical–obtaining.htm (accessed 20 August 2014)
2 Ref HS41, HSE Enforcement Policy Statement
3 See the statistical information from the HSE on their website www.hse.gov.uk/STATISTICS/tables/index.htm (accessed 20 August 2014)
4 See here the HSE statistical reports for 2005–2006 at www.hse.gov.uk/statistics/overall/hssh0506.pdf (accessed 20 August 2014)
5 See Evershed's health and safety update 2013–2014
6 See 'Corporate manslaughter: where's the proof?' by Eversheds staff, in *Safety and Health at Work* May 2013

7 See the TUC Report 'Health and safety: time for change, directors duties, the need for action'

8 'A survey of the use and effectiveness of the Company Director Disqualification Act 1986 as a legal sanction against directors convicted of health and safety offences' HSE Report 2007

9 See the HSE consultation on alternative penalties at www.hse.gov.uk/consult/condocs/penalties.htm (accessed 20 August 2014)

10 See here the HSE statistics 2005–2006 and the statistics on www.hse.gov.uk/STATISTICS/tables/index.htm (accessed 20 August 2014)

11 See here Paul Verrico 'Money back guarantee' *SHP* February 2013

7 Civil remedies

Introduction

In this chapter I wish to look at the remedies for a civil action brought in respect of an accident at work. In most cases an action will be brought in the Tort of Negligence or for breach of statutory duty. However, if the action is brought in contract the same kind of damages will apply, unless there is a contract term specifically defining the level of damages. It is also possible in Contract Law to obtain damages for loss of profits, which may be of interest if the worker is self-employed. However, I only want to look at damages payable in respect of personal injury and death, and these are the ones claimable in tort.

There are two main remedies awarded in a civil action, the award of damages or the granting of an equitable order such as an injunction. While the latter may be useful in matters involving a breach of the environmental law, the award of damages is usually the only relevant remedy, in respect of an injury at work. Therefore, I will begin by looking at the rules for the assessment of damages.

The award of damages

Types of losses and damages

Damages are awarded by a judge in the County Court or High Court, and the aim of such damages in tort is to put the claimant into the position they would have been in had the accident not occurred in the first place. Damages can be awarded to the injured party directly if he survives the accident, or they can be claimed by his dependants in the event of the injuries being fatal.

While common law principles decide liability and **quantum** (i.e. the amount of compensation paid), the award of damages is also subject to important statutes. The *Law Reform (Miscellaneous Provisions) Act 1934* and the *Fatal Accidents Act 1976* are important in determining the **heads of the award** (i.e. categories of damage compensated) and who can claim, while the *Social Security Administration Act 1982* provides for the deduction of state benefits in certain situations. It should always be borne in mind that in most cases involving industrial injury the party making the payment is the insurer of the person responsible for the accident (or **tortfeasor**).

Pecuniary and non-pecuniary loss

Damages are designed to cover the losses of the claimant and these fall into two main types, pecuniary and non-pecuniary loss. Pecuniary losses are those that involve a definable sum, such as present or future earnings or cost of care, while non-pecuniary losses are those that are difficult to quantify, such as pain and suffering or loss of amenity, where the loss must be quantified by the courts.

Special and general damages

There are two types of damages claimable here, special damages and general damages. Special damages cover definable pecuniary losses, which can easily be calculated, and will include loss of earnings up to the time of trial, as well as the cost of any private medical treatment up to that time. These losses have to be proved by the claimant, usually by presenting a wage slip or invoice. In addition, there are general damages and these cover pecuniary losses that are more difficult to calculate, such as loss of future earnings, as well as all non-pecuniary losses. General damages do not need to be specifically pleaded and will be determined by the court at the trial.

Joint and several liability

Where a number of tortfeasors are held to be jointly and **severally liable**, the claimant can obtain the full damages from any one of the defendants. In such a case the party who pays out the damages will be able to recover a contribution from the other parties, by way of the provisions of the *Civil Liability (Contribution) Act 1978*. Sometimes a claim may be may made against an uninsured party such as an employee, as per *Lister v Romford Ice & Cold Storage Ltd (1957)*, in which case the employer is unlikely to pursue a claim for a contribution, although this is possible in law.

If the employer goes into liquidation or is declared bankrupt it will still be possible to claim on the insurance, so long as the employer was insured at the time when the injury occurred. This is of particular importance in the case of latent diseases such as asbestosis, where the employer may have ceased trading many years before the injured party was aware that he had suffered injury. However, if the employer was not insured at the time of the accident there is no civil liability in respect of the failure to insure, only criminal liability as established in *Richardson v Pitt-Stanley (1995)*. It is not clear why civil liability should not extend to the employer as an individual in such a situation, but this was not provided for in the *Employers' Liability (Compulsory Insurance) Act 1969*.

Damages for personal injury

Introduction

Whether a person is injured in an accident at work or suffers long term damage due to a work-related illness, compensation may be claimed from the employer. Since 1966, nearly all such cases have been decided by a judge alone, rather than by a jury. As a result, the courts have been able to develop a common level of damages, as well as a common policy on awarding damages. This is sometimes referred to as the 'tariff' and details of the level of damages can be found in publications such as those published by the Judicial Studies Board.[1]

As detailed above, special damages will be awarded for specific pecuniary losses that can be accounted for, such as loss of earnings up to the trial and the cost of medical treatment. However, general damages will cover a wide range of pecuniary and non-pecuniary losses.

Pecuniary loss

Some pecuniary losses can be easily calculated and will usually be evidenced by way of a wage slip or an invoice. However, future losses must be assessed by the court and various methods have been devised to achieve this.

CALCULATION OF THE LOSSES

In the case of pecuniary losses, the court usually awards a lump sum, which is calculated to include future expectations of earnings. This is done by using a 'multiplier', which is a defined number of years and the 'multiplicand', which is the net annual losses, based on the rate of earnings at the time of trial, less any deductions for tax purposes. The multiplier is determined by an assessment of the number of years that the claimant would have been likely to work had he not been injured, but it is then scaled down to allow for the fact that the claimant benefits from the payment being made immediately, instead of spread out over a number of years. As a result, the multiplier will almost never exceed 18 years, even for the youngest victim. The money is expected to be invested in order to provide an income, and the courts assess this on the basis of an expected rate of return, which can be varied over the years.

The key problem with a fixed award is that the health condition of the claimant may vary over the following years, especially if there are unexpected complications with the injury. As a result, it is now possible for a claimant to obtain periodic payments to be paid over the course of the claimant's life. These payments can be varied by the court under the provision of the *Damages (Variation of Periodic Payments) Order 2005*. However, such awards pose a problem for the defendant as they create an uncertain liability over time, which may suddenly increase.

LOSS OF FUTURE EARNINGS

The main type of loss that is compensated is loss of future earnings. In calculating this, the court must take into consideration the life expectancy of the claimant using actuarial tables. In addition, the court should try to take into consideration promotion prospects or the possibility of redundancy. This can be done by varying the multiplier, so as to reduce the level of the overall award.

THE LOST YEARS

Where the claimant's life expectancy is reduced by the accident, it may be seen as unfair to simply calculate the loss of future earnings on the number of years he is likely to survive. As a result, the courts can now award compensation for the years during which he would have been expected to be earning. This is compensation for the so-called 'lost years' and has been allowed by the courts since the case of *Pickett v British Rail Engineering Ltd (1980)*. This is a speculative sum that may be easy to calculate for an older worker with a secure job, but could be very problematical for a younger worker.

LOSS OF EARNING CAPACITY

As an alternative to loss of earnings, the court can award a sum of money to compensate for loss of earning capacity. This will be the case where the claimant is still able to work but his ability to earn is reduced or he is more likely to be made redundant. This will include a situation in which the claimant is unable to obtain the necessary qualifications as a result of the accident. The assessment of damages here is more speculative and it must be shown that the claimant was really at a disadvantage as a result of his injuries. So in *Doyle v Wallace (1998)* it was necessary to show that the claimant's injuries were likely to prevent her obtaining the necessary qualifications to obtain promotion.

OTHER PECUNIARY LOSSES

Finally, it is possible to obtain damages to cover a number of other pecuniary losses under general damages, including the loss of any pension rights or the loss of use of a company car. The injured party can claim for the cost of future medical treatment or care that is seen as reasonable, as well as the purchase of any specialised equipment required as a result of the injury. There is no requirement that the claimant uses the National Health Service and so a claim for private treatment is acceptable, as was the case in *Harris v Brights Asphalt Contractors (1953)*. It is also possible to claim for the wages of a relative or friend who leaves work to look after him, as per *Donnelly v Joyce (1973)*, but this is not the case if that person is also the tortfeasor, as made clear in *Hunt v Severs (1994)*.

Non-pecuniary losses

The main non-pecuniary losses are for pain and suffering, and these have to be calculated by the courts. This type of loss includes both physical and mental suffering and will include an appreciation that the claimant's life expectancy has been shortened. However, no claim is possible where the claimant is unaware of the pain and suffering because he is in a vegetative state as a result of the accident, which was the case in *Wise* v *Kaye (1962)*.

In addition, there is also an award for loss of amenity such as physical movement or mental ability. It is possible to claim under this head even where the claimant is in a vegetative state, as was the case in *West & Son* v *Shepherd (1964)*, but the amount claimable will usually only be nominal.

Deductions from damages

The court will deduct from the award certain sums that the claimant may receive as a result of the injury. As regards private benefits received, any payment made to the claimant can be set off against loss of earnings, such as contractual sick pay and redundancy payments, unless they are voluntary payments from a disaster fund or a close friend, payments from a private insurance policy or private pension payments.

Under the provisions of the *Social Security (Recovery of Benefits) Act 1997*, payments over £2,500 received from the state for up to five years after the accident can be deducted from the loss claimable and paid directly to the state. This will cover a large number of state benefits, such as statutory sick pay, invalidity pension, severe disablement allowance and income support, and no account is taken of any reduction in the award due to contributory negligence. As a result, if a claimant has been claiming state benefits for some time following the accident, it is possible that they will receive no money at all, particularly if their award has been substantially reduced due to contributory negligence.

Damages for fatal injury

Loss of support

When a claim is brought on behalf of a claimant who has died, the rules are rather different. The relevant law applying here is based on two key statutes, namely, the *Law Reform (Miscellaneous Provisions) Act 1934* and the *Fatal Accidents Act 1976*.

Under the *Law Reform (Miscellaneous Provisions) Act 1934* it is possible for a claim to be brought in respect of pain and suffering, loss of expectation of life and loss of earnings up to the time of death. Although the claim is brought on behalf of the deceased, the damages will usually be paid into the estate for the benefit of the dependents of the deceased. A claim under the above heads is only possible where there is a period of time between the injury and the death of the victim, as the claim is in respect of the injury to the deceased. Brief periods of time between injury and death may be discounted as being part of the death itself, as made clear in *Hicks* v *Chief Constable of South Yorkshire Police (1992)*, a case arising from the Hillsborough disaster.

Under the *Fatal Accidents Act 1976* it is possible for the dependents of the deceased to claim for the loss of support from the deceased. The category of parties who are able to make such a claim is very wide and will include not only the spouse or former spouse, but also common law spouses, a parent or other **ascendant relative** (i.e. a grandparent), a child or other **descendant relative** (i.e. grandchildren), as well as a sibling or an aunt or uncle. The claim can only be made if the deceased had a valid claim in tort against the defendant, and any such claim could be reduced by contributory negligence under Section 5 of the *Fatal Accidents Act 1976* if the deceased was partly to blame for the accident.

Under this area of law, the dependents can claim for loss of support and the court will use the multiplier to calculate the amount of lost income. The claim can be made in respect of anticipated earnings (less the living costs of the deceased) where there is a real likelihood that they would have materialised, as made clear in *Taff Vale Railway* v *Jenkins (1913)*. The court will assess the prospects of the dependents as well as the deceased, as it will be expected that the dependents will mitigate their loss by seeking paid employment.

Other losses

In addition to a claim for loss of support, there are two other heads of claim. These include a claim for funeral expenses, as well a bereavement loss. The latter is now fixed at a standard sum of £10,000 and is only paid in respect of the death of a minor. Payment may go to either the spouse (if already married) or to parents of the deceased.

Deductions from benefits

Under Section 4 of the *Fatal Accidents Act 1976* any state benefits paid to the deceased between the injury and the time of death can be disregarded in calculating the damages due. This is in marked contrast to the position for claims for personal injury, when most benefits are deductable. Note that the damages awarded can be reduced if there was contributory negligence on the part of the dependents themselves, as it is they who will be obtaining the financial benefits.

Other types of remedy

Equitable remedies

Apart from an award of damages, it is possible for a claimant to obtain other remedies in civil law actions. In particular, a claimant can claim an injunction, which can be either mandatory or prohibitive in nature. A mandatory injunction requires something to be done, while a prohibitive one seeks to prevent an action. These remedies are typically used where there is a claim for environmental damage and the claimant wants to prevent the defendant from continuing to carry out the polluting activities.

Such remedies are awarded under **equity law** and as such they are made at the discretion of the court. As a result, the court can award damages if it thinks that this is a suitable alternative. In any event, the courts are unwilling to award any remedy that requires constant supervision.

Settlement out of court

We should also note that the vast majority of civil actions brought for injury at work are actually settled out of court. This is because the practice of payment into court puts considerable pressure on the claimant to settle once a payment is offered by the defendant. Should the claimant fail to match this amount when awarded damages, he will have to pay the legal costs of the defendant for all legal work carried out after the payment was offered. This can often run to many thousands of pounds and may exceed the amount of the award itself.

State benefits

In addition to the compensation claimable in law, it may be possible for an injured person to obtain payments from the state. The main payments are made under the industrial injuries scheme and will include compensation for prescribed industrial disease as well as injuries at work. In addition, there are a number of benefits that may be claimable, including disablement benefit and constant attendance allowance.

Conclusion

The remedies available in civil actions for Negligence are by their nature very limited in scope, and this may pose a problem to the claimant. The award of damages is usually by way of a fixed amount of money, but this can be paid to the claimant many years after the occurrence of the injury, resulting in the claimant having to live for several years on state benefits before receiving the award. Where liability is accepted and the issue at trial is merely the quantum of damages, it is sometimes possible to obtain some payments in advance by way of interim damages. However, such payments are often substantially reduced as a result of the recouping of social security payments.

The main problem with the system of civil remedies in tort is the high cost of administering the system. Although the social security system pays out a much larger sum of money, its administrative costs are a fraction of the tort system.[2] The adversarial system, with its demands for excessive documentation and opportunities for delay, does not always serve the claimant effectively or compassionately. Despite the reforms to the litigation process brought about by Lord Woolf, the civil remedy is still very cumbersome, time consuming and extremely expensive, and as such, in much need of reform.

Notes

1   See 'Guidelines for the assessment of general damages in personal injury cases' (2002) (6th edition) published by the Judicial Studies Board
2   See Harpwood 'Modern tort law', p. 422

8 · Conclusion

We can see from the previous chapters how the interaction of Civil and Criminal Law has played such an important part in the development of health and safety law. The matters that give rise to civil liability will often give rise to criminal liability as well, so issues such as the definition of what constitutes 'work equipment' are relevant both for the criminal prosecution and the civil claim. However, as we have seen, the aims of civil and criminal liability are quite different, the one focussing on deterrence and the other on compensation, so it may be problematic to use a similar definition.

Similarly, the nature of liability can be quite confusing. While criminal liability is often limited by the defence of what is 'reasonably practicable', civil liability is usually limited by what is 'reasonably foreseeable'. These legal concepts are very similar, but the key difference is that the concept of 'reasonably practicable' also involves a consideration of the cost in terms of time, money and convenience, which is absent from 'reasonably foreseeable'. In reality, the civil concept of 'reasonably foreseeable' does actually involve such factors, although not explicitly stated, as they are often taken into consideration in determining the extent of the liability.

In Criminal Law we have seen a tendency towards imposing greater liability on the defendant. The reversing of the burden of proof under Section 40 of the *Health and Safety at Work Act 1974* has led towards a result-based liability. In effect, the mere occurrence of an accident becomes the prima facie evidence of a breach of health and safety law, which the employer will find difficult to disprove, especially where the cause of the accident is difficult to determine, as can be seen in the case of *R v Chargot (2008)*.

In addition, we should note the widening scope of criminal liability here, in *HSWA*, especially where the breach of health and safety law leads to a fatality. Under Section 2 the employer is vicariously liable for the actions of his employees, even where they deliberately disobey orders, because this is seen as evidence of poor management supervision, as in *R v Gateway Foodmarkets (1997)*. Only where the employer can prove that he has a robust safety management system is he able to avoid liability following a fatality, such as in the case of *R v Hatton Traffic Management (2006)*.

Moreover, the liability of the employer extends beyond his employees to the actions of subcontractors working in his undertaking, by virtue of Section 3 of *HSWA*, as illustrated by the case of *R* v *Associated Octel Co. Ltd (1995)*.[1] This is very important because a criminal conviction can make it difficult for an employer to obtain contracts from prospective clients, who themselves fear that they may incur such liability if they engage an organisation with a poor safety record. In addition, a criminal conviction may make it more difficult for an employer to renew his employers' liability insurance, or in retaining accreditation from trade associations or health and safety organisations. The widening of the scope of liability here poses key issues for an employer who will need to exercise effective controls over the actions of subcontractors. The increased use of permits to work and the rigorous monitoring of subcontractors will be of increasing importance as employers seek to avoid such criminal liability.

The introduction of the *Corporate Manslaughter and Corporate Homicide Act 2007* is to be welcomed in as much as it deals with the problems highlighted by the unsuccessful prosecutions for Corporate Manslaughter in the past. On the other hand, liability here is only on the corporate body and personal liability for common law manslaughter is still difficult to prove unless we are dealing with a small organisation. Moreover, where prosecutions for Corporate Manslaughter are brought in tandem with individual manslaughter prosecutions, there is evidence that some employers are willing to accept a Corporate Manslaughter conviction in order to protect their directors from being imprisoned under the individual liability.[2]

Given the above, it is not surprising to find that there is such a high conviction rate for health and safety prosecutions, as outlined in the chapter on enforcement. Yet we need to remember that the purpose of Criminal Law is to deter bad safety practice, not to penalise an organisation because there has been an accident. More to the point, criminal prosecutions are often taken in the case of fatalities, but less often in the case of serious injuries, and yet the employer may be more at fault in the case of a serious injury compared to a fatality and will only face a prosecution under *HSWA*. Perhaps we should consider whether employers should be liable for such offences as exist in ordinary Criminal Law, such as causing grievous bodily harm or causing actual bodily harm.[3]

Arguably the main concern here is that the penalties for breach of the law are inadequate, with fines still quite low in relation to the turnover of many larger employers. While the introduction of custodial sentences for breach of the *HSWA* and other subordinate safety legislation has been a major development, the reality is that very few employers actually end up in prison. It is not clear how much of an incentive the present penalties are to some organisations to improve their performance and it might be useful to look at other possible methods of punishment as proposed by the HSE themselves.[4]

As regards civil liability, this has also been in a state of flux. The widening scope of liability in Negligence has been noted, with the development of liability for mental injury as opposed to purely physical injury, in respect of both mental breakdown and nervous shock. In addition, the effect of the decision in *Fairchild* v *Glenhaven Funeral Service (2002)* and the *Compensation Act 2006* has been to modify the rule on causation so as to allow mesothelioma victims, among others, to obtain proper redress. The rise of health-related claims here has been the major phenomenon of recent times and this is set to continue as the boundaries of medical knowledge expand, while increasing longevity means that more people are likely to be affected by diseases with a long term gestation. In addition, the continuing move towards a service-based economy and the growth of automation in dangerous industries has helped to reduce the incidence of traumatic injury at work.

However, in the last few years there has been a reaction against increasing liability on the part of employers and insurance companies. Employers have sought to transfer liability to subcontractors and even employees, although the courts have usually resisted this. On the other hand, we have seen an increasing move away from stricter liability to a more fault-based liability, especially with the ending of the right to claim for breach of statutory duty. Although future cases will determine how much effect this has, it is likely to remove the liability of employers for what might be seen as 'freak accidents', shifting the loss onto the employee. Given that the purpose of civil liability is to obtain compensation for injury, and that the employer is in the best position to insure against this, it is questionable whether these changes are really positive in nature.

Finally, we should note the impact of alternative forms of liability. Contractual liability has not really been developed, as the Tort of Negligence and breach of statutory duty provided an effective remedy. With the demise of the latter, it is possible that the contractual remedy may be used more often, and where there is a written contract this may even provide a form of strict liability. Equally so, we should not dismiss the Tort of Nuisance here, as liability could be developed to claim for loss of profits. This will be of importance if there is a growth in the number of labour-only subcontractors who are paid on a self-employed basis.

The impact of European Law on health and safety will continue to develop as a reaction to changing social and political attitudes. In this respect, the EU has already reacted by developing health and safety legislation to fulfil its commitment to improving the health and safety of workers under Article 153 of the Treaty of Rome. However, there have been strong counter pressures to this approach, especially in the UK, where there is a belief among some that economic development is being undermined by excessive regulation. The UK has resisted EU safety legislation, challenging the introduction of the Working Time Directive and securing a waiver procedure, as well as unsuccessfully opposing the introduction of safety representatives in non unionised workplaces. The recent campaign to leave the EU is to some extent driven by the desire to avoid the impact of such health and safety regulations, despite the fact that it was the UK that was initially the driving force behind the development of health and safety legislation in the EU. However, it is unlikely that the EU will continue to allow the UK to avoid European health and safety legislation and yet still obtain unrestricted access to EU markets.

Ultimately, we need to take account of the different roles of Criminal Law and Civil Law. Criminal Law is really concerned with the cause of accidents, and so we should only impose criminal liability where the evidence is that the employer was at fault in failing to ensure a safe working environment. As such, the focus here should be on the provision of an effective safety management system. Liability should not be determined by whether the management failure led to a fatality or merely a minor injury. Instead, we should be concerned with how effectively the employer managed health and safety, so as to avoid accidents in the first place. An employer should not be liable because there is an accident, but because there has been a serious management failure. A purely result-based liability undermines the purpose of criminal liability, as well as reducing respect for the law.

On the other hand, Civil Law is all about the consequences of an accident or disease. As such, we need to be ready to find a remedy for the injured party, even if this means that the liability is effectively strict in nature. In this way the insurance system can spread the cost onto the broadest shoulders instead of burdening the employee, who may be in no position to bear the costs. It may even be worth considering whether we should have a form of no-fault liability, as was suggested by the Pearson Report many years ago.[5]

We should recognise that there is always a balance to be drawn between economic efficiency and the needs of health and safety, and this issue will no doubt be contested in parliament and the courts over the next few years. However, we should not underestimate the immense benefits that improved health and safety has brought to the UK over the past century. In 2012, the UK constructed its Olympic Games complex without a single fatality; while in Quatar, the number of fatalities for migrant workers has become an international scandal. Health and safety is to a large extent a reflection of the values of a society, and the UK has a record to be proud of here.

Notes

1 See David Branson 'The burden of liability' *SHP* June 2007
2 See the case of R v Lion Steel Equipment Ltd (unreported) as discussed in Antrobus S. 'The criminal liability of directors for health and safety breaches and manslaughter' *Criminal Law Review* 2013, 4, 309–322
3 See David Bergman 'Corporate misconduct' (1999) *New Law Journal* 1849
4 See the HSE consultation on alternative penalties at www.hse.gov.uk/consult/condocs/penalties.htm (accessed 20 August 2014)
5 The Royal Commission on Civil Liability and Compensation for Personal Injury (1978) Cmnd 7054 (Pearson Report)

Appendix: key cases

Introduction

In this section I have set out some of the key cases for health and safety law that the student or practitioner needs to consider. These cases are likely to be modified by future judicial decisions, but most of them will probably stand the test of time. I have used modern terminology in all the cases, so that the term 'plaintiff' is replaced by the more modern term 'claimant'. The cases are organised in alphabetical order so that they are easy to look up and I have used the case name quoted in most references.

Adsett v K & L Steelfounders & Engineers Ltd (1953) 3 All ER 320

Facts of the case

The claimant was an employee in the defendant's foundry, where he contracted **pneumoconiosis**. He claimed that the employer was in breach of Section 47 of the *Factories Act 1937*, which required that 'all practicable measures' be taken to protect employees against any injurious dust. He argued that the defendant had failed to comply with the requirements of the statute and he had suffered injury as a result.

Decision of the court

The employer was not in breach of the statute as he had installed a dust extraction system as soon as he was aware of how to do this. The duty was complied with if the work was done as soon as it was technically possible, which was the case here.

Comment

This case illustrates the definition of the term 'practicable'. It is a higher level of liability than 'reasonably practicable', because it depends on technical knowledge, as opposed to 'reasonably practicable', which takes into consideration issues of time, cost and convenience. However, it is still a lower level than strict liability, which was what the employee was effectively claiming.

Armour v Skeen (1977) IRLR 310

Facts of the case

Following the death of a worker while working on a bridge over the River Clyde, the HSE prosecuted his employer, Strathclyde Regional Council, under Section 2 of the *Health and Safety at Work Act 1974*. They also prosecuted Mr Armour, the Director of Roads for the council, under Section 37(1) of the act, on the basis that the offence had been committed due to 'neglect' on his part as a director of the organisation. Mr Armour appealed against the conviction on the grounds that he was under no personal duty to carry out the statutory obligations of the council.

Decision of the court

The conviction of Mr Armour was upheld. The duty on the council to provide a safe system of work did not absolve Mr Armour of his personal duty of care to ensure that this was done. Although the offences under Section 2 were committed by the council as a corporate body, this was due to his neglect in failing to adequately supervise their activities. Although not technically a 'director' in a corporate sense, he was still a 'manager' and so was personally liable.

Comment

This was an early case under the *Health and Safety at Work Act 1974* and helped to define the nature of the personal liability under Section 37. Liability here falls on those persons who are the effective 'directing mind' of the organisation, in other words the person with effective control of the organisation.

It should be noted that liability does not attach to any employee designated as a 'manager', as this would impose too wide a liability. So in *R v Boal (1992)* the assistant manager of a bookshop was found not guilty of offences under the *Fire Precautions Act 1971*, as it was clear that he had no real control over the running of the premises and so did not fall within the definition of the term 'manager' in respect of Section 37.

Baker v Quantum Clothing Ltd (2011) UKSC 17

Facts of the case

The case was brought as a test case by a group of women exposed to noise levels of between 80 and 90 decibels (averaged over an 8 hour shift), working in the knitting industry during the 1970s and 1980s. They claimed that their employers owed them a common law duty of care to protect their hearing by providing them with ear protectors. Previous case law had held that this duty of care had only arisen where the noise exposure was above 90 decibels, even though there was an awareness of the damage of exposure to noise level above 85 decibels by the 1970s.

At first instance the court dismissed the claim, but on appeal to the Court of Appeal their claim was upheld. The decision was based on the fact that Section 29 of the *Factories Act 1961* required the employers to ensure that the place of work was 'safe' and that the liability here was strict as per *Larner v British Steel (1993)*. As the employers were aware of the danger of noise levels above 85 decibels from the1970s, liability for the injury was deemed to accrue from 1978. The appeal was to the Supreme Court.

Decision of the court

The appeal was upheld and the court re-imposed the original finding in favour of the defendant employers. Although there was information available about the risk of noise levels above 85 decibels, there was no legal obligation to provide protection above 90 decibels until the introduction of the *Noise at Work Regulations 1989*. Although some employers were aware of the risks of noise exposure below that level prior to 1989, it was wrong to impose a common law duty of care on the defendants, as they did not have the same level of information.

As regards the statutory liability under Section 29 of the *Factories Act 1961*, the court held that it only related to safety matters and not health issues such as noise. Moreover, it dismissed the idea that Section 29 imposed 'strict liability'. Instead the duty was qualified by the term 'reasonably practicable' and this is itself based on what is reasonably foreseeable. What Section 29 did was to shift the burden of proof, so that if there was a claim that the workplace was unsafe, the employer would have to prove that it was not reasonably practicable to have taken any other steps to deal with the problem.

Comment

The case highlights the fact that liability under statutory duties is not always strict but often qualified. It also indicates a narrower interpretation of the term 'safe' by the court, which limits it to non health matters.

Barker v Corus (2006) UKHL 20

Facts of the case

This was a case arising from a claim for mesothelioma against the defendant employer. Following the case of *Fairchild* v *Glenhaven Funeral Services (2002)* (see below) the courts had amended the normal rules of causation in such cases, as it was not scientifically possible for causation to be proved against any one defendant with such a disease (the so-called 'Fairchild' exception). As a result, all the possible defendants would be held 'jointly and severally' liable in respect of the whole claim and could then claim a contribution from any of the other parties if necessary.

In this case there was only one defendant still in existence, so he could not claim a contribution against any other party. In addition, the claimant himself was contributory negligent and his claim had been reduced by 20 per cent at the initial trial, where the defendant had been found jointly and severally liable under the Fairchild exception. The Court of Appeal had upheld liability on the basis of the Fairchild exception and allowed the deduction for contributory negligence. The defendant appealed on the basis that the Fairchild exception should not apply in such a case, because the defendant was only partly to blame.

Decision of the court

The House of Lords found that the original defendants were severally liable for the damage, and not jointly and severally liable. This meant that the damages should be apportioned between the parties and no one party could be sued for all the damages. As the claimant had only worked for the defendant for part of the time, this reduced the remaining defendant's liability to the claimant.

The court held that the Fairchild exception was a development of the earlier case of *McGhee* v *NCB (1973)* (see below).This imposed liability where the defendant's actions had 'materially increased' the risk of contracting dermatitis. However, this case would not apply where there were multiple causes, including where the claimant was partly responsible for the onset of the disease.

Where liability was joint and several, it was because the damage was seen as 'indivisible'. However, in this case, the 'risk' of the disease was increased by the defendant's actions, and as such the risk was 'divisible' and related to the period of exposure for which the defendant was responsible. Liability should therefore be several, not joint and several, so that the defendant was only liable for that amount of the period of exposure for which he was responsible.

Lord Rogers dissented from the majority decision on the basis that the liability in *McGhee* was based on contracting the disease, not increasing the risk of contracting it. Moreover, the Fairchild exception was a policy decision designed to place the risk on the defendant and his insurance company, rather than the claimant.

Comment

This case highlighted the role of the court in imposing **policy-based decisions**. The key question was whether liability was to fall on the defendant or the claimant, rather than deciding the case on any objective basis of fairness. In the event, the decision of the House of Lords proved very controversial and the government moved to overturn it with the *Compensation Act 2006*, which re-imposed joint and several liability in all mesothelioma cases.

Bradford v Robinson Rentals (1967) 1 All ER 267

Facts of the case

The claimant was a radio engineer working for the defendant. He was asked to undertake a long journey in very cold conditions, driving a van with no heater. As a result he suffered from frostbite and claimed damages from his employer.

Decision of the court

The employer was liable for breach of his common law duty to provide safe work equipment. The van was work equipment, and given the nature of the forecast weather conditions it was reasonably foreseeable that the employee would suffer from frostbite or some form of exposure if there was no heater in the van. The failure to provide a heater, or to abort the journey, meant that the employer was liable.

Comment

This case demonstrates the wide interpretation of the term 'work equipment', as well as providing an example of the factors that contribute to what is 'reasonably foreseeable'. The wide interpretation of work equipment has been followed in later cases, such as *Knowles* v *Liverpool City Council (1993)* and more recently *Spencer Franks* v *Kellogg Brown Root Ltd (2008)* (see below). Also note that so long as it was possible to claim civil liability under the *Provision and Use of Work Equipment Regulations (1998) (PUWER)* the interpretation of the term 'work equipment' was very important, as *PUWER* imposed strict liability on the employer.

Cambridge Water Co Ltd v Eastern Counties Leather PLC (1994) 2 WLR 53

Facts of the case

The claimant company owned the right to drill for underground water in the local area. They discovered that the water had been contaminated by toxic chemicals deposited on the land by the defendant company. The claimant sued for liability under the rule in *Rylands* v *Fletcher (1868)*, claiming that the defendants were strictly liable for any escape of a harmful substance where the use of the land was 'non-natural'. The High Court held that the defendant was not liable as the use of the land was 'natural', but the Court of Appeal disagreed with this and held that the defendant was liable. The defendant appealed to the House of Lords.

Decision of the court

The defendant was not liable as the pollution caused by his waste material was not 'reasonably foreseeable'. Whether the use of his land was natural or non-natural was not important in determining liability.

Comment

Although this was an action for damage to property under the rule in *Rylands* v *Fletcher (1868)*, it shows how the law here is approximating to the principles in common law Negligence, in so much as liability is not strict but limited by what is 'reasonably foreseeable'.

Close v Steel Company of Wales (1961) 2 All ER 953

Facts of the case

The claimant was injured by an electric pedestal drill when the drill bit shattered and a piece hit him in the eye. It was accepted that there was no liability in Negligence, because the accident was not 'reasonably foreseeable'. However, the claimant argued that there was a breach of statutory duty, as the drill was not adequately guarded in accordance with Section 14(1) of the *Factories Act 1937*, which imposed strict liability.

Decision of the court

The Court of Appeal held that Section 14(1) of the *Factories Act 1937* only required a guard to prevent an employee obtaining access to dangerous moving parts, not to prevent such parts being ejected from the machine and striking the employee. Moreover, Section 14(1) only required the guarding of dangerous parts of machinery, and parts were not deemed to be dangerous unless it was reasonably foreseeable that they could cause injury, which was not the case here.

Comment

This case revealed the limitations of the *Factories Act 1937* in providing protection for employees. As a result, when it was replaced with the *Provision and Use of Work Equipment Regulations 1998*, the opportunity was taken to extend the protection to cover such eventualities, as made clear in Regulation 12(3)(d) of the same. The liability here is strict and does not require the risk of injury to be 'reasonably foreseeable'.

Corn v Weir Glass (Hanley) Ltd (1960) 2 All ER 300

Facts of the case

The claimant was injured when he slipped while carrying a sheet of glass down a stairway. The stairway was not provided with a handrail and the claimant sued for breach of Regulation 27(1) of the *Building (Safety, Health and Welfare) Regulations 1948*, which required such a handrail to be provided.

Decision of the court

The claimant failed in his action because the court held that he would not have used the handrail even if provided, as he was using both hands to carry the sheet of glass. Therefore the failure of the employer to comply with the regulations was not the cause of the accident.

Comment

This case illustrates the fact that to claim for breach of statutory duty it is necessary to show both a breach of the relevant statute and that the breach was the cause of the injury. It is the case that volenti on the part of the claimant, or the actions of a third party, can break the chain of causation and negate liability on the part of the employer. See here the later case of *Mc Williams v Sir William Arrol & Co (1962)*.

Davie v New Merton Board Mills Ltd (1959) 1 All ER 346 HL

Facts of the case

The claimant employee was injured while using a new chisel supplied by his employer. The defect in the chisel was not reasonably detectable by the employer, who believed the work equipment to be of sound quality. The claimant argued that the employer was in breach of his common law duty of care in Negligence.

Decision of the court

The employer was not liable as the defect was undetectable. Any action would lie against the manufacturer, and this had to be taken by the claimant himself under common law Negligence.

Comment

The legal position established here was reversed by the passing of the *Employers' Liability (Defective Equipment) Act 1969*, which made the employer primarily liable for any defective work equipment. In addition, there is also statutory liability under *PUWER*, which was interpreted strictly as in *Stark* v *Post Office (2000)* (see below). However, this statutory duty was abolished with effect from October 2013.

Donaldson v Hays Distribution Services Ltd (2005) SLT 733

Facts of the case

The claimant was a customer at a shopping centre who was injured by a lorry at the defendant's premises. She argued that the defendant was liable under Regulation 17 of the *Workplace (Health, Safety and Welfare) Regulations 1992 (WHSWR)* in respect of ensuring the safety of pedestrians on traffic routes. The defendant held that the relevant legislation only covered persons at work, while the claimant argued that the term 'pedestrians' applied to workers and non-workers.

Decision of the court

The case went to appeal at the Scottish Inner House. The court held that the legislation did not apply to non-workers and any liability in respect of the claimant would have to be brought in Negligence or under the *Occupier Liability Act 1957 (OLA)*.

Comment

The claimant had sought to use *WHSWR* because they imposed strict liability on the defendants. However, such protection was not provided for non employees, so she would be obliged to use common law Negligence or take an action under *OLA 1957*, both of which imposed fault-based liability.

Donoghue v Stevenson (1932) AC 562

Facts of the case

The claimant suffered from severe gastro-enteritis when she drank a bottle of ginger beer manufactured by the defendant company, in which there was a decomposed snail. The action was ultimately appealed to the House of Lords on the question of whether there could be liability to a person who did not have a contract with the defendant.

Decision of the court

There was a duty of care imposed on the defendant because the claimant could not have examined the product and because there was a duty on the defendant to look after the interests of persons affected by his actions. This was the so-called 'neighbour principle' as set out by Lord Atkin, who argued that a person had to take reasonable care to avoid acts or omissions that he could reasonably foresee would be likely to injure his neighbour. His 'neighbour' was defined as being a party he could reasonably foresee, as being likely to be affected by his act or omissions.

Comment

This case was the starting point for most consumer based negligence actions. It also laid down the legal principle that liability in Negligence was based on what was 'reasonably foreseeable' and this principle underpins the whole of Negligence law.

Edwards v National Coal Board (1949) 1 All ER 743 CA

Facts of the case

The claimant's husband was killed by a roof fall in a mine due to the failure of the defendant to shore up the pit props in the area in which he was working. The defendant argued that it was not cost effective to shore up all the props and so the defendant was not liable for his death.

Decision of the court

The Court of Appeal held that the defendant was liable, as he did not need to shore up all the pit props, only those where the risk was greatest. This could have been determined by a risk assessment and the total cost would not have been excessive compared to the risk.

Comment

This case was the leading case in defining the nature of the term 'reasonably practicable'. Lord Asquith defined the term as consisting of a balance between the risk on the one hand and the cost in terms of time, trouble and money to rectify the situation. The assessment of what was reasonably practicable had to be determined on the basis of the facts available prior to any accident. It is clear that this is a narrower definition than the term 'practicable', as it involves taking into consideration the cost of dealing with the risk.

Ellis v Bristol City Council (2007) EWCA Civ 685

Facts of the case

The claimant slipped on a floor at a care home run by the defendant council, suffering injury. She argued that the defendant was in breach of Regulation 12(1) of the *Workplace (Health, Safety and Work) Regulations 1992*, which imposes a strict liability on employers to ensure that any floor in a workplace is of such a construction as to be 'suitable' for the purpose for which it is used. The defendant argued that Regulation 12(3) of the regulations applied, in respect of keeping the floor clear of any substance that might cause a person to slip, and that this duty was qualified by what was 'reasonably practicable'. The County Court held that Regulation 12(3) applied and that the defendant was not liable because he had done everything 'reasonably practicable' to avoid the accident.

Decision of the court

The Court of Appeal overturned the County Court decision, holding that liability arose under Regulation 12(1), as this regulation applied when the problem occurred frequently, which was the case here; whereas Regulation 12(3) only applied where the spillage was an unusual occurrence.

Comment

In preferring to apply Regulation 12(1) the court was effectively imposing a strict liability requirement, not one that was fault-based. However, this case illustrates the difficulty in distinguishing between liability under 12(1), in respect of floor construction, and Regulation 12(3), in respect of maintenance issues. This issue also arose in the earlier case of *Lewis v Avidan (2005)*, where a sudden water leak was seen as falling within Regulation 12(3).

Fairchild v Glenhaven Funeral Services (2002) UKHL 22

Facts of the case

The claimant had been exposed to asbestos fibres while working for the defendant and other employers, and as a result had contracted mesothelioma, a fatal lung disease. Due to the nature of the disease it was not possible to determine which of the employers was responsible for the claimant contracting the disease, which can be caused by the inhalation of a single asbestos fibre.

The High Court and Court of Appeal had held that the claimant could not obtain a remedy, as it was not possible to prove causation against a particular employer. In common law Negligence the damage has to be proved on the balance of probabilities and this was not possible against any one defendant.

Decision of the court

Reversing the decision of the lower courts, the House of Lords held that in these exceptional circumstances the defendants were 'jointly and severally' liable and the damages would be apportioned in accordance with the period of time that the claimant had worked for a particular employer. In this respect, they followed the decision in *McGhee* v *NCB (1973)*, holding that liability arose where the defendant had made a 'material' contribution to the risk of contracting the disease.

Comment

This is a key decision in Negligence, which modifies the normal rule of causation for this type of case. It was a clear policy decision, determined by the practical difficulties of proving causation, as well as a belief that it would be 'deeply offensive' if the claimant was denied any remedy in the circumstances. The so-called 'Fairchild' exception has been limited so far to mesothelioma cases. See also here the case of *Barker* v *Corus (2006)* above.

Ferguson v John Dawson & Partners (Contractors) Ltd (1976) 3 All ER 817

Facts of the case

The claimant was a bricklayer who worked on a building site as a self-employed labour-only subcontractor, popularly known as 'lump labour'. He was injured in a fall at work and claimed for breach of the *Construction (Working Places) Regulations 1966*, which required the claimant to show that he was an employee. In the hearing at first instance it was held he was self-employed, and so ineligible to claim under the regulations. He appealed to the Court of Appeal.

Decision of the court

Although the contractual relationship was one of self-employment, the court preferred to look to the reality of the situation rather than the form.

The claimant was under the total control of the defendant and was an integral part of the defendant's organisation. As a result, the court held that he was effectively an employee and so was able to claim.

Comment

This case established the principle that the court will not allow the formal contractual relationship to limit the liability of the employer under health and safety law. The use of self-employment has expanded rapidly in recent years, mainly to avoid the effects of employment protection law. The court was concerned to ensure that this should not be used to dilute the protection under health and safety law; otherwise much of the law would have become unenforceable. However, it is to be noted that most recent subordinate legislation uses the term 'worker' and not 'employee', and this will cover self-employed persons. A similar decision to the above was made in the later case of *Lane* v *Shire Roofing (1995)*.

Fytche v Wincanton Logistics PLC (2008) UKHL 31

Facts of the case

The claimant was a lorry driver employed by the defendant. His lorry became stuck in heavy snow and he spent three hours digging it free. During this time, freezing water leaked into his boots and he later suffered from frost-bite. The claimant said that the safety boots he was provided with were in breach of Regulation 7(1) of the *Personal Protective Equipment at Work Regulations 1992*, which requires all personal protective equipment (PPE) to be maintained in 'an efficient state, in efficient working order and in good repair'. The defendant argued that the boots were not provided for walking outside in the snow, but to protect him from goods falling on his feet.

Decision of the court

On appeal to the House of Lords, it was held that the employer was not liable. The PPE was suitable for the purpose for which it was provided, which was to protect against injury to the feet from falling goods. The PPE was not provided to protect the claimant's feet against inclement weather.

Comment

This is one of many cases brought under either the *Provision and Use of Work Equipment Regulations 1998* or the *Personal Protective Equipment at Work Regulations 1992*. This is because both regulations impose strict liability on the employer for the provision and maintenance of such equipment. However, the courts will not allow unlimited liability here; the equipment need only be suitable for the purpose for which it is supplied, not for any purpose for which it could be used.

General Cleaning Contractors Ltd v Christmas (1953) 2 All ER 1110 HL

Facts of the case

The claimant was a window cleaner employed by the defendant. The company used a method of cleaning windows that involved standing on the window ledge and holding onto the window frame. While doing this, the window sash fell and the claimant fell to the ground suffering injury. He claimed for damages in common law Negligence.

Decision of the court

The method of work used was unsafe and had caused the claimant's injury. As a result the employer was liable. The employer should have devised a safer system of work to ensure the safety of the claimant.

Comment

This case made it clear that the responsibility lay with the employer to devise a safe system of work for his employees. This requirement is now made clear in Section 2(2)(a) of the *Health and Safety at Work Act 1974*.

Haseldine v Daw & Sons Ltd (1941) 3 All ER 156

Facts of the case

The defendant occupier of premises engaged a firm of lift engineers to repair a lift on his premises. The claimant was killed when the lift failed, suddenly falling to the bottom of the lift shaft. The claimant's relatives brought a case on his behalf to claim damages, alleging negligence on the part of the occupier of the premises.

Decision of the court

The occupier was not liable, as he had engaged what appeared to be a competent contractor. Instead the liability fell on the contractor for any loss and he could be sued directly by the claimants.

Comment

This case established that a defendant is not liable for the actions of a contractor, whom he had engaged in the reasonable belief that the contractor was competent. This is of importance where a member of the public is injured by a contractor working on the defendant's premises. Liability can be shifted to the contractor, so long as it can be proved that the defendant took all reasonable care to ensure that the contractor was competent. This would involve checking his qualifications and references. Where an employee is injured, a claim would be made initially against the employer, but the employer could then join such a contractor as a co-defendant and ensure that any liability was borne by that party.

Johnstone v Bloomsbury AHA (1991) 2 All ER 293

Facts of the case

The claimant was a junior doctor who was required by his contract to work 40 hours a week and also to be on call for a further 48 hours a week. He argued that his hours of work were far too long and he was suffering from stress and anxiety, which could lead to him making mistakes that could harm patients. He wanted the court to issue a declaration that it was unlawful to require him to work so many hours in a week.

Decision of the court

The Court of Appeal held that the claimant had a case to argue. In particular they agreed that there was an implied term in the contract of employment, that the employer ensured that the employee's health was not put in danger. Moreover, the term that required him to work up to 88 hours was arguably void as being unreasonable under the *Unfair Contract Terms Act 1977*, which prevents an employer excluding or limiting his liability for Negligence leading to personal injury.

Comment

This case illustrates the fact that long working hours are seen as a health hazard and the employer must ensure that the health of their employees is not jeopardised by requiring him to work excessive hours. The *Working Time Regulations 1998* provide a restriction on hours worked for most employees, but the above case provides a useful protection for those employees who have exempted themselves from its requirements.

Knowles v Liverpool City Council (1993) IRLR 6

Facts of the case

The claimant was injured while laying pavements for the defendant council, when one of the paving stones broke due to it being defective. The claimant sued his employer for Negligence, arguing that he was liable for defective work equipment under the *Employers' Liability (Defective Equipment) Act 1969*. The employer argued that the flagstone was a work material, not work equipment.

Decision of the court

It was held by the court that the flagstone was work equipment. The court was unwilling to draw a distinction between work equipment and work materials, especially given that the worker used the flagstone to push other flagstones into position.

Comment

This case shows how difficult it is to distinguish between work equipment and work materials. The distinction is of great importance because the employers have greater liability for the former, both under the above legislation and of course under the *Provision and Use of Work Equipment Regulations 1998*.

Latimer v AEC Ltd (1953) 2 All ER 449 HL

Facts of the case

The claimant was injured when he slipped on a patch of oil in the defendant's factory. The oil had been mixed with water, which had leaked through the roof of the factory in a heavy rainstorm. Although some sawdust had been put down to try and deal with the problem, not all of the floor had been covered and the employee slipped on an area not so treated.

Decision of the court

The employers were not liable, as it was not reasonable in the circumstances to have prevented any injury. A reasonably prudent employer would not have been expected to act differently, as it was not reasonable to have expected him to close the whole factory.

Comment

This case appears to limit the liability of the employer for the safety of his employees by defining what is expected of the 'reasonable employer'. However, it is questionable whether such a view would be taken today, given that the employer now has to provide insurance under the *Employers' Liability (Compulsory Insurance) Act 1969*. It is likely that the employer would now be liable, as the workplace was clearly unsafe, and this was not a freak accident but was 'reasonably foreseeable'. Note that the relevant criminal duty under Regulation 12(3) of the *Workplace (Health Safety and Welfare) Regulations 1992* is qualified by what is 'reasonably practicable' and this will take into consideration the practical problems and cost implications of closing the premises, which is not explicitly taken into consideration in the civil liability.

Lister v Romford Ice & Cold Storage Co. Ltd (1957) AC 555

Facts of the case

The claimant was an employee of the defendant and was injured by a van driven negligently by his son, another employee. The employer paid compensation to the claimant, but sought to recover an indemnity from the son.

Decision of the court

Although the employer was vicariously liable to the claimant for his injuries, the employer was entitled to seek a contribution from the son, especially given that the two lived together in the same household.

Comment

This case established the right of the employer to seek a contribution from a negligent employee, where he has to compensate another employee because of that negligent person's actions. This right is not usually exercised because it is likely to cause problems within the workplace, but it may be exercised in cases of gross Negligence, especially if the insurance company seek to recoup their losses.

McFarlane v EE Caledonian (1994) 2 All ER 1

Facts of the case

The claimant was a crew member on a rescue vessel attending the fire on the Piper Alpha oil rig, where 167 workers lost their lives. He claimed for psychiatric injury caused by his experiences as a result of witnessing the loss of life. The defendant stated that the claimant was never in personal danger and could not claim for nervous shock as a 'secondary victim' as there was no relationship of close love and affection with the deceased persons. The employer also disputed the argument that he could claim as a rescuer, under the provision of *Chadwick v British Transport Commissioners (1967)*.The initial court held that the defendant was liable for psychiatric injury.

Decision of the court

The Court of Appeal reversed the decision and held that the claim would fail. It was clear that the claimant was not a secondary victim. If he was to argue that he was a primary victim, the court held that he would have to prove that he was in personal danger, had reason to believe he was in such danger or that he came to the danger willingly as a rescuer. The court held that none of these situations applied here as the claimant was never in danger and no person of normal fortitude would have believed he was in danger. As regards the argument that he was a rescuer, this definition was restricted to persons who were directly involved in the rescue operation and that did not include the claimant.

Comment

This case restricts the categories of persons who can claim as 'secondary victims' for psychiatric injury as a result of experiencing a horrific accident. The courts are concerned to avoid opening the 'floodgates' to excessive claims, so there is no general liability in respect of experiencing such horrors. Instead liability is limited to persons who can show close love and affection with the victims as per the definition in *Alcock* v *Chief Constable of South Yorkshire Police (1992)*. Alternatively, the claimant must play a direct part in the rescue operations.

McGhee v National Coal Board (1971) 1 WLR 1

Facts of the case

The claimant was employed in a brickworks run by the defendant. He was unable to get a shower at the end of the shift as there were no washing facilities, and he had to travel home with brick dust on his skin. He claimed that this led to him contracting dermatitis, so he was claiming for damages caused by the defendant's failure to provide such washing facilities as required by law. The initial court and the Court of Appeal held for the defendant.

Decision of the court

The House of Lords reversed the decision and supported the claim. The employer was liable because the lack of washing facilities 'materially increased' the risk of the claimant contracting the disease.

Comment

This case clarifies the rules on causation and decided that where an act or omission makes a 'material contribution' to the risk of contracting a disease, that is sufficient grounds for liability. This decision was relied upon in the case of *Fairchild* v *Glenhaven Funeral Services (2002)* (see above), and it would appear to widen the scope of causation so as to make a party liable, even where there is no clear proof that his acts or omissions caused the disease.

Marshall v Gotham Co Ltd (1954) AC 360; (1954) 1 All ER 937

Facts of the case

An employee of the defendant company was killed by a roof fall in the defendant's mine. The fall had been caused by an unusual geological condition known as 'slickenside', which had not been seen in the mine for over 20 years. The claimant was the widow of the deceased, who brought an action for common law Negligence and breach of the employer's statutory duty under the *Metalliferous Mines General Regulations 1938*. The Negligence action was dismissed, but the breach of statutory duty was allowed at first instance. In the Court of Appeal the claim was dismissed on the basis that the statutory duty was qualified by the term 'reasonably practicable' and that the employer had complied with that requirement. The claimant appealed to the House of Lords on a point of law as to the meaning of the term 'reasonably practicable'.

Decision of the court

Dismissing the appeal, the House of Lords held that the employer was not liable, as the term 'reasonably practicable' is not based on what was practicable as a matter of engineering possibility, but on the basis of whether the time, trouble and expense involved appeared disproportionate to the risk of injury.

Comment

The case helped to define the legal principle of 'reasonably practicable' and built on the concept laid down in *Edwards* v *the National Coal Board (1947)* (see above). The standard is a lesser one than 'practicable', as it takes into consideration issues of cost, in terms of time, trouble and convenience, which are not factors in determining what is 'practicable'.

Mersey Docks and Harbour Board v Coggins and Griffiths (1946) 2 All ER 345

Facts of the case

The claimant was a crane driver employed by Mersey Dock and Harbour Board. He was hired out to the defendant, another firm of stevedores, who were given the power to instruct him as to what jobs were to be done. The crane driver injured a third party who sued for Negligence and the question arose as to which employer was vicariously liable for his actions.

Decision of the court

The employer who hired out the crane driver was responsible for the employee's actions, in other words, as the employer, Mersey Dock and Harbour Board had ultimate control over his actions.

Comments

This case clarifies the rules on vicarious liability. The party hiring out the employee retains responsibility for the actions of his employee, even though the party to whom he is hired has immediate control over what is done. The position may be different if the employee himself is injured. In that case, the party who has hired the employee may be responsible for the injury, especially if the employee is relatively unskilled, as in the case of *Garrard* v *A.E Southey & Co Ltd (1952)*.

Paige v Freight Hire (Tank Haulage) Ltd (1981) IRLR 13 EAT; (1981) 1 All ER 394

Facts of the case

The claimant was a female employee of the defendant company. She was dismissed from her job as a Heavy Goods Vehicle driver when the goods she was required to transport were deemed to be harmful to unborn children, so that women of child bearing age were prohibited from transporting such goods. The claimant argued that this was a breach of the *Sex Discrimination Act 1975*.

Decision of the court

The Employment Appeal Tribunal held that the dismissal was fair, as it fell within Section 51(1) of the *Sex Discrimination Act 1975*, which allowed the law to be overruled if it was contradicted by another piece of legislation. In this case, Section 2(1) of the *Health and Safety at Work Act 1974* required the employer to prevent the claimant being involved in such work, and no alternative employment was available at the firm.

Comment

This is the leading case that upholds the priority of health and safety issues over discrimination issues. As such, it has stood the test of time and so indicates that sex discrimination is no grounds to challenge health and safety restrictions that affect only one sex. It should be noted that the same situation exists for discrimination on the grounds of race or disability, see here *Panesar* v *Nestle Co Ltd (1980)*.

Pannett v McGuiness (1972) 3 All ER 157

Facts of the case

The claimant was a five–year-old child who was injured when he fell into a fire on a building site. The fire was being supervised for most of the time, but the attendant had gone off for a break, leaving the fire unattended.

Decision of the court

The defendant was liable for failing to take sufficient steps to prevent injury to the claimant. The premises should have been fenced in or the fire should have been supervised at all times.

Comment

This case shows how the liability to young children is higher than to older children or adults. Arguably the parents should have some responsibility here for failing to supervise a child of such a young age, see here *Phipps* v *Rochester Corporation (1955)*.

Paris v Stepney Borough Council (1951) 1 All ER 42 HL

Facts of the case

The claimant was employed by the defendant company as a welder. He had only one good eye, as he had lost the other in the war. The claimant was injured in his good eye by a flying metal chip when trying to remove rusty metal on the underside of a bus. The employer argued that it was not normal for employees to wear eye protection when carrying out such work.

Decision of the court

The employer was liable for injury to the claimant. Given the known disability of the claimant, a reasonable employer should have ensured that eye protection was worn when undertaking such a job.

Comment

This case establishes the principle that a higher duty of care exists to employees who are disabled. A similar higher duty of care applies in the case of young persons or persons of limited mental capacity who may be employed.

R v Associated Octel Co. Ltd (1995) ICR 281 CA

Facts of the case

The defendant operated a chemical plant which was regulated by the *Control of Major Accident Hazard Regulations 1999*. This required the operator to use a permit to work system when maintenance work was being carried out. The defendant engaged a specialist contractor to repair a storage tank, but did not properly implement the permit system, with the result that there was an accident, and an employee of the contractor was injured. The defendant was convicted of a breach of Section 3(1) of the *Health and Safety at Work Act 1974 (HSWA)*. The defendant argued that he was not responsible for the failures of an independent specialist contractor.

Decision of the Court

Both the Court of Appeal and the House of Lords upheld the conviction. Under Section 3(1) of *HSWA*, the defendant was responsible for conducting his undertaking in a safe manner, and the carrying out of the maintenance activity by an independent contractor fell within that definition.

Comment

This case illustrates the wide interpretation given to Section 3(1) of *HSWA*, which means that a party may be liable for the failings of an independent contractor, if he is carrying our work on that party's site. The key issue here was the failure of the defendant to effectively operate the permit to work system. In addition, the fact that the defendant was aware of the hazards faced by the contractor, and the necessary controls required, was also important. However, an employer or occupier of premises will not be liable, if he does not have that specialist knowledge, and it is reasonable for him to depend on the expertise of the independent contractor, as made clear in *Haseldine* v *Daw (1941)* above.

R v British Steel (1995) 1 WLR 1356

Facts of the case

The defendant company engaged two subcontractors on a labour-only basis to re-position a 7.5 tonne steel platform at their Sheffield Works. The job was supervised by a section manager from the defendant company. Unfortunately the subcontractor's employees carried out the work negligently, causing the death of one of their employees. The defendant was convicted in the Crown Court for a breach of Section 3(1) of the *Health and Safety at Work Act 1974* because he failed to conduct his undertaking is such as way as to ensure, so far as is reasonably practicable, the health and safety of persons not in his employment. The defendant appealed on the basis that he should not be responsible for the actions of a junior member of staff and that he had done everything 'reasonably practicable' to avoid the accident. Reference was made to the case of *Tesco Supermarkets* v *Nattrass (1972)*, where the defendant company was not held liable, except for the actions of senior managers deemed to be the 'directing minds' of the organisation.

Decision of the court

Upholding the conviction, the Court of Appeal said that Section 3(1) of *HSWA* imposed absolute liability on the employer, subject to the qualification of what was 'reasonably practicable'. It was not possible for the defendant to delegate that duty to a manager. The case of *Tesco Supermarkets* v *Nattrass (1972)* was not relevant here as it applied to consumer law not health and safety law.

Comments

This case makes clear that liability for health and safety is absolute and non-delegable and that liability exists vicariously for the negligence of all employees, and arguably even non employees under the control of the employer. However, the Court of Appeal suggested that the HSE would not prosecute for all breaches by an employee, as this would lead to an absurd level of liability. It is clear that such a prosecution would be seen unfavourably by the courts; see here the decision in *R* v *Nelson Group Services (Maintenance) Ltd (1999)* below.

R v Chargot (2008) UKHL 73

Facts of the case

The defendant was a subcontractor who was engaged to carry out construction work on a farm. One of his employees, a dumper truck driver, was killed when his truck overturned. There were no witnesses to the accident and it was not clear why the accident had happened. The truck driver was not wearing his seatbelt at the time. The defendant was convicted under Section 2(1) of the *Health and Safety at Work Act 1974 (HSWA)* for failing to ensure, so far as is reasonably practicable, the health and safety of his employees. The defendant held that the jury had been wrongly advised, that liability would exist if the prosecution could simply identify and prove a risk of injury. Instead, he claimed that the prosecution should have had to prove a particular act or omission on his part. The conviction was upheld in the Court of Appeal and this was further appealed to the House of Lords on a point of law.

Decision of the court

Upholding the conviction, the House of Lords held that the liability in Section 2 and 3 of *HSWA* was 'result-based'. Under Section 40, the burden of proof switches to the defendant once a prima facie breach is proved and it is up to the defendant to show that he did everything that was 'reasonably practicable' to ensure health and safety. It was not up to the prosecution to prove an act or omission by the defendant, only that the end result of ensuring health and safety had not been achieved.

Comments

This case underlines the fact that liability under Section 2 and 3 of *HSWA 1974* is likely to be proved where there is an accident, even if its cause cannot be determined. While liability is not absolute, it is certainly relatively strict, and this is different from most criminal liability where guilt must be proved beyond reasonable doubt. Note here that liability under Section 7 of *HSWA* is different as it is based on a duty to take reasonable care rather than ensuring an end result, see here *R v Beckingham (2006)*.

R v Gateway Foodmarkets (1997) IRLR 189

Facts of the case

An employee at one of the defendant company's stores was killed on entering a lift control room, when he fell down an open trapdoor. The employer had issued instructions prohibiting store staff from entering this area, but regular access had occurred to deal with a particular problem with the lift. The employer argued that he should not be liable for the unauthorised action of junior staff of which he was unaware. He appealed against his conviction under Section 2 of *HSWA* in the Crown Court.

Decision of the court

The Court of Appeal upheld the conviction. The employer was liable for the actions of his employees because he failed to control what was happening at a local level. The court followed the decision in *R v Associated Octel (1994)* and rejected any idea that the liability of the organisation was restricted to the actions of senior managers.

Comments

As with *R v British Steel (1995)* above, the duties under Section 2 and 3 cannot be delegated. The courts are no doubt concerned that an employer may try to distance himself from the unauthorised actions of lower level employees when these are so often the reasons for an accident in the first place. As with *R v Associated Octel (1994)*, liability arose here because the employer failed to effectively supervise operations.

R v Hatton Traffic Management (2006) EWCA Crim 1156

Facts of the case

The employer was engaged in carrying out road repairs on the A66 at night. Two employees were electrocuted when they moved a mobile lighting tower and it struck an overhead power cable. In the prosecution in the Crown Court, the defendant argued that he had done everything reasonably practicable to avoid the accident and said that it would not have happened if employees had followed the company safety instructions.

The prosecution argued that what was 'reasonable foreseeable' to the employer, was not relevant to the question of liability and that Regulation 21 of the *Management of Health and Safety at Work Regulations 1999 (MHSWR)* precluded the defendant relying on the act or default of an employee as a defence. The judge ruled that evidence of what was foreseeable was relevant and that Regulation 21 of the *MHSWR* did not preclude the defendant from relying on the act or default of an employee as a defence. The prosecution appealed to the Court of Appeal against the judge's ruling.

Decision of the court

The Court of Appeal held that the defendant was able to submit evidence of the likelihood of risk and that he could put forward evidence to show that the accident was the fault of one or both employees. They held that the term 'reasonably practicable' was not a defence as such, but qualified the duty of care itself under the *Health and Safety at Work Act 1974*.

Comment

This case clarifies the nature of the term 'reasonably practicable', which is not a defence as such, but a qualification of the duty of care. It also makes it clear that the liability is not absolute but qualified and that an employer will not always be liable simply because there is an accident. If he can show that the employee deliberately ignored instructions and it was not reasonable for the employer to prevent it, then liability will be negated.

R v Nelson Group Services (Maintenance) Ltd (1999) IRLR 646

Facts of the case

The employer installed, maintained and serviced gas appliances, and employed a large number of gas fitters, who were trained to carry out the job safely. One of the fitters left a gas appliance in an unsafe condition, exposing the occupier of the premises to a risk to their health and safety. As a result, the defendant was convicted in the Crown Court of a breach of Section 3(1) of the *Health and Safety at Work Act 1974*. The defendant argued that he had done everything 'reasonably practicable' to avoid a breach of the law and that the negligent employee was alone responsible.

Decision of the court

The Crown Court conviction was quashed by the Court of Appeal. It was held that an employer could establish a defence of 'reasonably practicable', even where an employee had acted negligently. In determining whether the employer did all that was 'reasonably practicable', the court would take into consideration relevant safe systems of work available and the training provided to employees.

Comment

The case established the defence that an employer was not always liable for the negligence of his employees. This defence was later qualified by the passing of Regulation 21 of the *Management of Health and Safety at Work Regulations 1999*.

R v P (2007) EWCA Crim 1937 CA

Facts of the case

The case arose out of the death of a six–year-old boy who was thrown out of a fork lift truck on which he was being given a ride at the defendant's docks. The fork lift truck was hit by another truck, which was carrying an unclamped newspaper reel, a practice known as 'riding reel', which was dangerous and unauthorised. The Managing Director of the company was prosecuted for breach of Section 37(1) of the *Health and Safety at Work Act 1974 (HSWA)*, which makes a senior manager individually liable if a corporation commits a breach of *HSWA* due to that manager's connivance, consent or neglect. The trial judge stated that in order to prove neglect, it was necessary for the prosecution to prove that the defendant knew of the unsafe practices going on at the dockside. The prosecution appealed against this preliminary ruling to the Court of Appeal.

Decision of the court

On appeal, it was held that the ruling was incorrect. It was only necessary for the prosecution to show that the defendant should have been aware of the facts, not that he was actually aware of them.

Comments

This decision makes it clear that the term 'neglect' in Section 37(1) of *HSWA* includes not only turning a 'blind eye' to the situation, but also failing to check what was actually happening in the workplace.

R v Porter (2008) EWCA Crim 127

Facts of the case

The defendant was the Head Teacher of a primary school where a young child died after falling down a short flight of steps in the school playground. He was convicted of a breach of Section 3(1) of the *Health and Safety at Work Act 1974 (HSWA)* and he appealed against the conviction.

Decision of court

The Court of Appeal upheld his appeal against conviction. The risk against which the defendant had to take precautions had to be real, not fanciful or hypothetical. In this case, the risk of injury was part of the risk of everyday life and not the kind of risk contemplated by *HSWA*.

Comment

The court will decide whether the risk is real as opposed to fanciful or hypothetical, on the basis of evidence such as the occurrence of previous accidents and the likely actions of the persons involved. In this case, the evidence suggested there was no real risk. The occurrence of an accident does not mean that the risk was a real one. Note that to some extent this case qualifies the decision in *R v Chargot (2008)* discussed above.

R v Swan Hunter Shipbuilders Ltd (1981) IRLR 403

Facts of the case

The case arose out of an accident at the defendant's shipyard, where a contractor's employee failed to properly disconnect an oxygen hose at the end of a shift, allowing the creation of an oxygen rich atmosphere overnight. When hot work recommenced the next day there was an explosion and fire, which killed eight Swan Hunter employees. The defendant was prosecuted under Section 2(2) of the *Health and Safety at Work Act 1974 (HSWA)* for failing to provide adequate information on health and safety. The defendant appealed against the conviction on the grounds that the failing was on the part of the contractor.

Decision of the court

Upholding the conviction, the Court of Appeal held that the defendant had endangered the lives of his own employees, because he had failed to provide information on safe working practices to the employees of the contractor. This failure ultimately put his own employees' lives in danger.

Comment

The case illustrates the fact that to protect your own employees, it may be necessary to ensure that third parties are also properly informed of any potential hazards involved. It makes clear the wide nature of liability under Section 2 of *HSWA*.

Robb v Salamis (2006) UKHL 56

Facts of the case

The claimant was employed by the defendant on an oil rig at sea. He was injured when the rungs of a bunk-bed came away suddenly, causing the claimant to fall from the upper bunk. He claimed for breach of Regulations 4 and 20 of the *Provision and Use of Work Equipment Regulations 1998 (PUWER)*. His action for damages in the Sheriff Court failed, as did the appeal to the Court of Session. He then appealed to the House of Lords on a matter of law as to whether the regulation applied.

Decision of the court

Although the claimant was off-duty at the time, given the nature of the job the bunk-beds were still work equipment as accepted by the Court of Session. The liability under Regulation 4 of *PUWER* required the employer to provide 'suitable' work equipment, and in this respect 'suitable' meant in any respect that it is 'reasonably foreseeable' will affect the health and safety of any person. In determining what was 'reasonably foreseeable' the court held that the employer had to anticipate problems and not just wait for them to happen. As the ladders were clearly unsafe in their design, the employer was liable.

Comment

This case is important because it helps define the term 'work equipment'. It is clear that the term applies to anything used at work, whether directly or indirectly. The concept of what is 'reasonably foreseeable' is interpreted widely, so as to require the employer to anticipate possible hazards. However, a specific accident does not need to be foreseeable, only the general nature of the danger involved.

Rose v Plenty (1976) 1 All ER 97

Facts of the case

The case arose out of an accident to a boy of 13 who was injured while helping a milkman to deliver his round. The defendant employer specifically prohibited the employment of children, but the practice still went on. The defendant argued that he was not vicariously liable for the accident, because the employee had deliberately ignored work instructions and so was not acting in the 'course of employment'.

Decision of the court

The employer was vicariously liable, despite the deliberate flouting of his instructions. The definition of the term 'course of employment' included any action designed to benefit the employer, which was the case here.

Comment

This case indicates the wide definition of the term 'course of employment'. The courts have long been unwilling to restrict it to actions authorised by the employer, as to do so would mean that many accidents would fall out of its scope and the employer would not be liable. As a result, vicarious liability will extend even to grossly stupid actions by an employee that were expressly prohibited, such as lighting a cigarette while delivering petrol to a petrol station, see here *Century Insurance* v *Northern Ireland Transport Board (1942)*.

Rylands v Fletcher (1868) LR 3 HL 330

Facts of the case

The defendant employed independent contractors to build a reservoir on his land in order to provide water power for his mill. When excavating the land, the contractors discovered the remains of old mineshafts, which connected up to the mine workings operated by the claimant. The contractors failed to effectively block up the mineshafts so that, when the reservoir was filled, water flooded into the mine workings of the claimant causing damage. The defendant was unaware of the failure of the contractors to effectively block the mineshafts, so could not be held liable in Negligence. In addition, he was not vicariously liable for the actions of the contractors, as they were not his employees and the law had not yet developed to consider the position of independent contractors.

Decision of the court

The defendant was found strictly liable for the escape of the water, on the basis of a new legal principle laid down in the case. This imposed strict liability on a person bringing anything onto his land that could do mischief if it escaped. The rule was held only to apply to non-natural use of the land.

Comment

The definition of non-natural use has been subject to much interpretation in later years. Increasingly, many industrial activities are seen as falling outside of the definition, including the provision of piped water as in *Transco PLC* v *Stockport Metropolitan B.C. (2004)*. In addition, the strict nature of the liability was modified by the later decision of *Cambridge Water Co.* v *Eastern Counties Leather (1994)*, which has limited the losses claimable to those that are 'reasonably foreseeable', in much the same way as they are in common law Negligence.

Smith v Crossley Brothers (1951) 95 Sol Jo 655 CA

Facts of the case

The claimant was seriously injured by a practical joke played on him at the apprentice training school run by the employer. The defendant claimed it was due to wilful behaviour and he could not be liable for such actions, as they were outside of the 'course of employment'.

Decision of the court

This was a wilful act and the employer was not liable, as he could not have foreseen that such an action would occur.

Comment

The definition of 'course of employment' does not extend to wilful acts unrelated to the business of the employer. This was also the case in *Coddington* v *International Harvester of GB (1969)*. However, there is evidence that the claimant was not a willing participant in the activity and it is arguable that he would nowadays be able to claim. The more recent case of *Lister* v *Hesley Hall (2002)*, which involved child abuse in a children's home, would appear to support such a view.

Smith v Northamptonshire County Council (2009) UKHL 27

Facts of the case

The claimant was employed by the defendant council as a care assistant and driver. Her duties included collecting elderly people from their homes and taking them to a day centre. One of the elderly people was a wheelchair user who had a ramp installed by the NHS at her front door to enable her to get out of the house. When the claimant was helping her one day, the ramp collapsed causing injury to the claimant's foot. The claimant sued for breach of the statutory duty under Regulation 5 of the *Provision and Use of Work Equipment Regulations 1998*, to provide and maintain work equipment. The lower court decided in her favour, but the Court of Appeal held that the ramp was not 'work equipment' and was not being used at work. The claimant appealed to the House of Lords on both of these points of interpretation.

Decision of the court

The appeal was dismissed by the House of Lords. While the ramp could be 'work equipment' in some circumstances, it was not so here because it was not adopted as part of the employer's business or undertaking. This would have required the ramp to have been provided by the employer or a third party and used with the employer's consent. The fact that the employer had regularly inspected it was not sufficient.

Comment

This is an interesting case because it suggests that the fact that an employer inspects equipment used by employees in the course of their employment does not make that equipment work equipment. There was a dissenting judgment, which held that liability should exist if the use of the equipment was known to the employer, as he could have instructed the employee not to use it if necessary. The case seems to limit the accepted scope of work equipment and will pose problems for employees who deal with members of the public. In either case, it is difficult to see why there was no liability in common law Negligence.

Spencer Franks v Kellogg Brown Root Ltd (2008) All ER (D) 26

Facts of the case

The claimant worked as a technician on an oil rig. He was injured while repairing a self-closing door on the rig when the closing mechanism sprung loose and hit him in the face. The claimant sued for breach of Regulation 4 of the *Provision and Use of Work Equipment Regulations 1998*, on the basis that the door was work equipment and was not 'suitable' for its purpose. The initial case in the Sheriff Court held the door was 'work equipment', but this was rejected on appeal to the Court of Session. The claimant appealed to the House of Lords on a point of law.

Decision of the court

The door and the closing mechanism were work equipment, as the term applied to all equipment on an offshore rig that was provided for use at work, even when it formed part of the fabric of the platform. The claimant succeeded in his appeal.

Comment

This case provides a wider definition of work equipment than previous cases. In particular it overrules the case of *Hammond* v *Commissioner of Police for the Metropolis (2004)*, which had restricted the definition of work equipment to equipment used by the claimant and not equipment worked upon by him. Indeed, the new definition is so wide that it encompasses the concept of the workplace as covered by the *Workplace (Health, Safety and Welfare) Regulations (1992)*. However, it is to be noted that this case related to an accident on an oil rig, and under the *Offshore Installations (Operational Safety, Health and Welfare) Regulations 1976* all parts of a rig are effectively deemed to be work equipment.

Stark v Post Office (2000) ICR 1013

Facts of the case

The claimant was a postman who was severely injured when the calliper of the bicycle he was using suddenly broke and he was catapulted over the handlebars. He brought a claim for Negligence as well as for breach of statutory duty under Regulation 6 of the *Provision and Use of Work Equipment Regulations 1992 (PUWER)*, which required that work equipment be maintained 'in an efficient working order and in good repair'. In the court of first instance it was held that no liability existed in Negligence or for breach of statutory duty as the accident was completely unforeseeable. The claimant appealed against the dismissal of the claim for the breach of statutory duty on the basis that the liability under Regulation 6 of *PUWER* was absolute.

Decision of the court

In the Court of Appeal it was held that the regulation did impose an absolute liability and so the claimant could maintain his appeal and obtain damages.

Comment

This is a very important case, which confirmed the absolute liability of the statutory duty under what was then Regulation 6 of *PUWER*. In particular, it should be noted that the defendant was liable, even though the accident could not be foreseen, so that liability was not fault-based as in Negligence, but absolute. As a result, it could be seen that an action under the statutory liability was much more effective as a means of securing compensation, even in the case of a so-called 'freak accident'. It was concerns over this case on the part of employers' organisations, which led to the abolition of the right to claim compensation for breach of statutory duty, ending a right that had existed for over a hundred years.

Stokes v Guest Keen & Nettlefold (Bolts & Nuts) Ltd (1968) 1 WLR 1776

Facts of the case

The claimant was the widow of an employee who had died of scrotal cancer, caused by excessive exposure to cutting oil, which had soaked into his work overalls. The defendant company had not warned its employees about the dangers of such oils, even though the Factories Inspectorate had made the risk known to employers.

Decision of the court

The employer was liable for the death of the claimant's husband. He had breached the duty of care in Negligence because he should have been aware of the problem and taken steps to bring it to the attention of his employees.

Comment

This case defined the standard of care expected in Negligence, which was the standard of a reasonable and prudent employer. As such, he was expected to ensure the safety of his employees in the light of the knowledge that he had, or ought to have had. This meant that the employer has to keep abreast of health and safety developments in his area of work.

Summers & Sons Ltd v Frost (1955) 1 All ER, HL

Facts of the case

The claimant was injured while using an abrasive wheel, when his thumb was caught in the gap between the wheel and the guard. He brought an action under Section 14(1) of the *Factories Act 1937*, which imposed an absolute duty to guard the dangerous parts of the equipment. The employer argued that the existing guarding was adequate.

Decision of the court

The House of Lords held that the guarding requirements were absolute and that the guarding had to provide complete protection, even for the careless or inattentive worker.

Comment

This was a very literal interpretation of the law and it meant that all abrasive wheels had to be totally enclosed, making them unusable. The position was rectified some 15 years later by the passing of the *Abrasive Wheels Regulations 1970* (now revoked). In the meantime, the Factories Inspectorate tended not to enforce this interpretation of Section 14.

Sutherland v Hatton (2002) EWCA Civ 76

Facts of the case

This was an appeal involving four different employers, in cases where the employees had ceased work due to psychiatric illness caused by work-related stress. The employers argued that they could not be liable unless the stress was clearly brought to their attention.

Decision of the court

The Court of Appeal upheld the appeals in three of the cases, holding that the employers could not be liable unless the injury was reasonably foreseeable. In most cases this will mean that the employee must bring the issue to the attention of his employer. Otherwise, an employer is entitled to assume that employees are able to deal with the normal pressures of the job. However, the mere fact that an employee continues to work on while under stress does not exonerate the employer from liability if the employee then suffers injury due to work-related stress, see here the decision in *Intel Corporation (UK) Ltd v Daw (2007)*.

Comments

This case clarified the nature of the decision in *Walker v Northumberland County Council (1995)* (see below) in that it put the onus on the employee to raise the issue of stress, unless the evidence of the stress was so obvious that a reasonable employer would recognise it and do something about it.

Threlfall v Hull City Council (2010) EWCA Civ 1147

Facts of the case

The claimant was employed by the defendant council as a street cleaner and part of his job involved picking up bags of rubbish from the gardens of unoccupied council houses. He sustained a serious cut to his hand when picking up one of the bags and argued that the defendants were in breach of Regulation 4 of the *Personal Protective Equipment at Work Regulations 1992 (PPE Regulations)*, in as much as the gloves provided were not 'suitable' for the task. In the lower courts it was held that the council was not liable, as the risk of injury was very low and so the gloves were appropriate for the task.

Decision of the court

The Court of Appeal held for the claimant on the basis that the gloves were not suitable for the task, as evidenced by the accident. Although a risk assessment had been carried out, it did not properly identify the possible risks. The gloves were not effective in preventing injury, and as such were unsuitable and in breach of Regulation 4 of the *PPE Regulations*.

Comment

The decision imposes a strict interpretation of the *PPE Regulations*, similar to that of Regulation 4 of *PUWER*. The suitability of the PPE should be determined by a risk assessment, which will assess if they are appropriate to the risk. However, the court held that the liability was effectively result-based, in as much as the occurrence of the accident demonstrated that the PPE was not effective. As such, the liability was strict and not based on common law concepts of what was 'reasonably foreseeable'.

Viasystems (Tyneside) Ltd v Thermal Transfer Ltd & Others (2005) EWCA Civ 1151

Facts of the case

The claimant was a factory owner whose premises had been damaged by the actions of the defendant, who was installing air conditioning on the premises. The damage was caused by the negligence of an employee of a labour-only subcontractor called C & T Metalwork Service, whose employee accidently set off a sprinkler system, flooding the claimant's premises. C & T Metalwork Services were subcontracted to a firm called S & P Darwell Ltd, who in turn were subcontracted to Thermal Transfer Ltd to carry out the work. The issue arose as to which of the contractors was liable for the actions of the negligent employee. At first instance, it was held that liability should lie solely with C & T Metalwork Services, on the grounds that he was vicariously liable for the negligence of his employee. C & T appealed on the grounds that the employee involved was effectively under the control of both himself and S & P Darwell, and so that liability should be shared.

Decision of the court

The Court of Appeal held that vicarious liability should be shared equally between both C & T Metalwork Services and Darwell, and damages should be apportioned under the *Civil Liability (Contribution) Act 1978*.

Comment

This case overturned the long established legal principle that vicarious liability could only apply to one party and could not be shared. The increasing use of subcontractors has meant that it is sometimes very difficult to establish who is in effective control, so the decision reflects the reality of modern subcontracting. It also addresses the problem of civil liability being shifted down the line of subcontractors to the smaller organisations, which may be less likely to have arranged effective insurance cover. It is also in line with the position in Criminal Law under Section 3 of the *Health and Safety at Work Act 1974*, where liability may be shared for the actions of an employee of a subcontractor, where the party in overall control has not provided effective supervision. See here the case of *R v Associated Octel (1995)*.

Walker v Northumberland County Council (1995) IRLR 35

Facts of the case

The claimant was employed as a social worker by the defendant council, dealing with issues of child abuse. His workload increased over time and he eventually had a nervous breakdown in 1986. He was promised extra support when he returned to work, but this was not forthcoming. As a result, he had a subsequent and more serious breakdown a few months after returning and he was unable to work again. He claimed damages from the defendant for causing his mental illness, due to the failure of the council to ensure that his workload was not excessive.

Decision of the court

Although the defendant was not liable for the initial breakdown, he was liable for the subsequent one, as he was aware of the problem and had failed to take action to deal with it. The defendant was in breach of his duty to provide a safe system of work.

Comment

This was the first case to establish the liability of the employer for mental injury caused by workplace stress. It was based on the failure of the employer to ensure a safe system of work and, as such, extends the concept to include psychological issues. The decision reflects the changing nature of work in the UK, with an increasing number of jobs now posing a risk of mental stress rather than physical injury. However, it also poses the problem that psychological injury may be more difficult to foresee, as much depends on the mental wellbeing of the claimant, which the employer may not be aware of. The decision was later clarified by the case of *Sutherland* v *Hatton (2002)*.

Wilsons & Clyde Coal v English (1937) 3 All ER 628 HL

Facts of the case

The claimant was a miner who was injured by a moving plant, when walking along a mine shaft at the end of a shift. He claimed damages for his injury, but the employer argued that he was not personally liable as he had appointed competent persons to operate the plant and the negligence had been on their part. The lower court found for the claimant, but the case was eventually sent to the House of Lords to decide on the above point of law.

Decision of the court

The employer was personally liable for the health and safety of his employees and could not delegate the responsibility to another person. The duty of care here was to provide competent staff, adequate materials and a safe system of work. The employer had failed to do this and so was liable.

Comment

This case **disapproved** of the concept of 'common employment' by which an employee was deemed to have consented to the risks of his fellow employees' Negligence. The court held that liability was personal to the defendant employer and could not be delegated in this way. Moreover, the case set down the basic elements of the duty of care in Negligence, which (with some modifications) became the basis of the statutory duty of care under Section 2(2) of the *Health and Safety at Work Act 1974*.

Glossary of legal and medical terms

Absolute liability Legal liability for which there is no defence available. It is different from strict liability where a defence may be possible.

Accused In a criminal case, the party who is charged with the offence.

Appellate court A court that only hears appeals from lower courts (i.e. the Supreme Court).

Asbestosis Potentially fatal lung disease caused by inhaling asbestos fibres; often caught by persons who carry out lagging of boilers.

Ascendant Relative Relatives from the previous generation (i.e. parents and grandparents).

Balance of probabilities Where the facts suggest more than a 50 per cent likelihood that the claimant is right in law.

Beyond all reasonable doubt Where it appears to be 95 per cent certain that the defendant is liable.

Body corporate An organisation that is a separate legal entity from its members, such as a public limited company.

Causation The link between the breach of the duty of care in law and the occurrence of the injury or loss to the claimant.

Claimant The person or organisation taking the legal action, either the injured party or a relative of the same. Previously referred to as the Plaintiff.

Contribution Where a defendant is able to recover part of the damages paid from another defendant who is also liable for the injury or loss.

Course of employment Any activity done in work time designed to benefit the employer, even where the action is carried out in a grossly negligent manner. The employer is vicariously liable for such actions if they cause loss or injury to a third party.

Court of first instance A court where the case is heard for the first time (i.e. the Magistrates' Court or Crown Court).

Damages   The money payable by the defendant to the claimant as compensation for causing injury or loss.

Defendant   The person or organisation against whom the legal action is taken. In criminal cases the term accused is also used.

Descendant relatives   Relatives from the next generation (i.e. children and grandchildren).

Directing mind   A person in an organisation who is the effective decision maker at a high level. This can be the owner, director or a senior manager. The organisation itself may be liable for gross negligence manslaughter if personal liability for the death can be attached to a person who is a 'directing mind' of the organisation.

Disapproved   Where a higher court refuses to overrule a previous decision or legal principle but states that it is not good law and should not be followed in future.

Due diligence   Taking all reasonable care. Sometimes used as a defence instead of 'reasonably practicable', as in the *Electricity at Work Regulations 1989*.

End user   The party that actually benefits from the work of the employee (i.e. the employer to whom an agency worker is loaned).

Equity law   Legal principles applied by the court, which are based on concepts of fairness and equity rather than strict legal rights. Courts can issue injunctions rather than award damages under this area of law.

Equitable remedy   This is a remedy that is not available as of right under law but is granted at the discretion of the court.

Estate   In terms of payments to a deceased person, the rights over property and other assets he possesses that will pass on to his beneficiaries under the will or his next of kin. As regards land, it involves the degree, quantity, nature and extent of interest that a person has over the land.

Exclusion or limitation clause   A clause that excludes or limits the liability of a party for breach of contract or Negligence. Such a clause may be written into a contract or brought to the attention of the contracting party before they enter into the contract.

Fault-based   Legal liability that requires fault to be proved on the part of the defendant.

Good faith   An action that is not motivated by malice.

Heads of the award   The categories of damage compensated in a civil action such as loss of earnings or pain and suffering.

Injunction   A court order by which a person is required to perform or is restrained from performing a particular act.

Joint and several liability   Where a group of defendants are held to be responsible collectively for the damages awarded as well as on an individual basis.

Judicial review   The reference of a decision of a court or a government body to the Queens Bench Division of the High Court, who will determine whether the decision was properly reached within the requirements of the law.

Latent injury   An injury that takes some time to manifest itself, such as a disease with a long term gestation (i.e. asbestosis).

Legal capacity   The ability to enter into a legal arrangement such as a contract. This is determined by such factors as age and mental capability.

Manslaughter   Criminal liability for causing the death of another person due to gross negligence or recklessness.

Mesothelioma   A fatal lung disease that may be caused by the inhalation of a single asbestos fibre.

Misrepresentation   An untrue statement of fact, made to another party, in order to induce him to enter into a contract with themselves or a third party.

Nervous shock   The legal term used to describe post traumatic stress disorder, a mental trauma caused by exposure to a sudden shock, such as witnessing the violent death of a colleague at work.

Pneumoconiosis   Potentially fatal lung disease caused by the inhalation of dust particles, especially coal dust.

Policy-based decision   Where the court decides to make a decision on the basis of what they think is socially fair or expedient, even if it contradicts long established legal principles. This was the case with the so called 'Fairchild exception' to the rules on causation, established in *Fairchild v Glenhaven Funeral Services (2002)*.

Point of law   A dispute over the meaning of the law (i.e. the definition of the term 'work equipment').

Prima facie   On the face of it; in other words, there appears to be a breach of the law unless there are other circumstances not obvious at the time.

Prosecution   The organisation responsible for bringing a criminal case against a defendant, usually the HSE or a Local Authority.

Quantum   The amount of the damages that the claimant believes should be awarded.

Quasi-criminal   Semi criminal, in other words the use of enforcement notices.

Quasi-legal Part of the legal system that is outside the normal court hierarchy, such as the tribunal system.

Remoteness Another term for legal causation. The defendant is only liable for consequences of the breach of the duty of care that are 'reasonably foreseeable'. He is not responsible for all possible consequences.

Res ipsa loquitur This means 'the facts speak for themselves'. This legal principle will apply where there is an accident that could only have come about by negligence on the part of someone and the defendant was at all times in control of the situation. In this case, the burden of proof is shifted from the claimant to the defendant.

Result-based Liability that is partly determined by consequences (i.e. liability for manslaughter requires a person to have been killed).

Severally liable Where individual defendants in a group of defendants are held to be only liable for that portion of the damages which is due to their particular negligence, as was the situation in *Barker* v *Corus (2006)*.

Silicosis Potentially fatal lung disease caused by inhaling silica dust, often caught by quarry workers and stonemasons.

Strict liability Liability in law for which fault does not need to be proved against the defendant. It differs from absolute liability in so far as a defence may still be possible.

Third party A person or organisation outside of the contractual relationship between employer and employee, such as a contractor or a lawful visitor.

Tort A legal wrong, such as Negligence or Nuisance, for which a remedy is available in law.

Tortfeasor The creator of the tort, in other words the defendant in a civil action for Negligence or Nuisance.

Tortuous action An action under one of the torts (i.e. Negligence or Nuisance).

Traumatic injury An injury caused by some immediate and usually violent event (i.e. a fall from height).

Undue influence Use of influence over a party, to cause him to enter into an agreement against his will.

Volenti The defence of consent, where a claimant is held totally responsible for the accident that has led to the legal action.

Warrant   Written document stating that the holder is a HSE Inspector, and as such entitled to exercise the powers accorded to him under Section 20 of the *Health and Safety at Work Act 1974*.

Zero hour contract   A contract by which the employer is engaged to work for the employer, but there is no guaranteed minimum number of hours per week provided.

Bibliography

Books

Harpwood, V. 'Modern tort law', 7th edition (2009) Routledge Cavendish, Oxford

Selwyn, N. 'The law of health and safety at work 2009/10' (2009) Croner, London

Slapper, G. 'Blood in the bank: social and legal aspects of death at work' (1999) Ashgate, Aldershot

Study book for the NEBOSH National Diploma: 'managing health and safety', 4th edition (2011) RMS Publishing, Stourbridge

'Tolley's Health and safety at work handbook', 22nd edition (2010) Lexis Nexis, London

Articles

Antrobus, S. 'The criminal liability of directors for health and safety breaches and manslaughter' *Criminal Law Review* 2013, 4, 309–322

Bergman, D. 'Corporate Misconduct' (1999) *New Law Journal* 1849

Branson, D. 'The burden of liability' *Safety and Health Practitioner* June 2007, 53–54

Branson, D. 'Temping fate' *Safety and Health Practitioner* February 2010, 30–32

Branson, D. 'Fault lines' *New Law Journal* 2013, 15–16

Bridges, K. 'All in good order' *Safety and Health Practitioner* 2005

Clark, C. 'Minority report' *Safety and Health Practitioner* October 2009, 53–54

Dobson, A. 'Shifting sands: multiple counts in prosecutions for Corporate Manslaughter' *Criminal Law Review* 2012, 3, 200–209

Evershed's staff 'Corporate manslaughter: where's the proof?' *Safety and Health at Work* May 2013

Field, S. and Jones, L. 'Death in the workplace: who pays the price?' *Company Law* 2011, 32(6), 166–173

Forlin, G. 'Developments in health and safety' *Archbold Review* 2011, 7, 7–8

Howes, V. 'Liability for breach of statutory duty – is there a coherent approach?' *Journal of Personal Injury Law* 2007, 1, 1–14

Howes, V. 'Duties and liabilities under the Health and Safety at Work Act 1974: a step forward?' *Industrial Law Journal* 2009, 38(3), 306

Howes, V. 'Who is responsible for health and safety of temporary workers? EU and UK perspectives' *European Labour Law Journal*, 2011, 2, 379–400

Hsalo, M. 'Abandonment of the doctrine of attribution in favour of gross negligence test in the Corporate Manslaughter and Corporate Homicide Act 2007' *Company Lawyer* 2009, 30(4), 110–112

Lee, M. 'Tort law and personal injury in the regulatory state' *Journal of Personal Injury Law* 2011, 3, 137–143

McCarthy, F. 'Case comment: Baker v Quantum Clothing Group Ltd' *Journal of Personal Injury Law* 2011, 3, C122–127

O'Doherty, S. 'Personal injury: causation: a floating concept' *New Law Journal* 2009, 159, 809

Spencer, J.R. 'Criminal liability for accidental death: back to the Middle Ages?' *Cambridge Law Journal* 2009, 68(2), 263–265

Tomkins, N. 'First principles in employer liability' *Journal of Personal Injury Law* 2010, 3, 131–138

Tomkins, N. 'The workplace regulations and strict liability' *Journal of Personal Injury Law* 2010, 2, 65–69

Tomkins, N. 'Case comment, personal injury: liability-health and safety at work' *Journal of Personal Injury Law* 2011, 1, C14–18

Tomkins, N. 'Is health and safety an "albatross"?' *Journal of Personal Injury Law* 2012, 2, 90–94

Tomkins N. 'Case comment, Berry v Star Autos Ltd and Others (QBD May 2012 unreported)', *Journal of Personal Injury Law* 2013, 2, C95–98

Vericco, P. 'Money back guarantee' *Safety and Health Practitioner* February 2013, 31–32

Weir, R. 'Personal injury: not aloud' *New Law Journal* 159, 1279

Government publications

'A survey of the use and effectiveness of the Company Director Disqualification Act 1986 as a legal sanction against directors convicted of health and safety offences' HSE Report (2007)

'Access to justice', Final report of the Right Honourable the Lord Woolf Master of the Rolls (July 1996)

'Guidelines for the assessment of general damages in personal injury cases' (2002) (6th edition) published by the Judicial Studies Board

Lofstedt, R. 'Reclaiming health and safety for all: an independent review of health and safety legislation' (November 2011) Cm 8219

Report of court No 8074: public inquiry into the sinking of the Herald of Free Enterprise chaired by Lord Justice Sheen (September 1987)

Robens, A. 'Safety and health at work: report of the Committee 1970–2' (1972) Cmnd 5034, London: HMSO

The Royal Commission on Civil Liability and Compensation for Personal Injury (1978) Cmnd 7054

TUC reports

'Health and safety: time for change, directors duties, the need for action' (2013), London: Trades Union Congress

Index